FUNDAMENTALS OF MODERN ALGEBRA

A Global Perspective

FUNDAMENTALS OF MODERN ALGEBRA

A Global Perspective

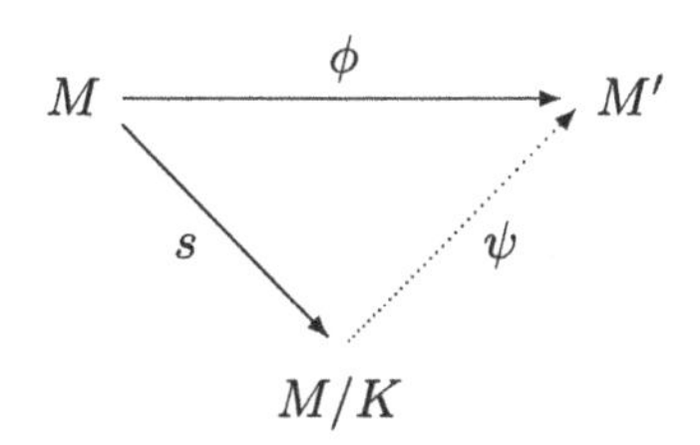

Robert G Underwood

Auburn University at Montgomery, USA

World Scientific

NEW JERSEY • LONDON • SINGAPORE • BEIJING • SHANGHAI • HONG KONG • TAIPEI • CHENNAI • TOKYO

Published by

World Scientific Publishing Co. Pte. Ltd.
5 Toh Tuck Link, Singapore 596224
USA office: 27 Warren Street, Suite 401-402, Hackensack, NJ 07601
UK office: 57 Shelton Street, Covent Garden, London WC2H 9HE

Library of Congress Cataloging-in-Publication Data
Names: Underwood, Robert G. (Robert Gene)
Title: Fundamentals of modern algebra : a global perspective / by Robert G. Underwood
(Auburn University at Montgomery, USA).
Description: New Jersey : World Scientific, 2016. |
Includes bibliographical references and index.
Identifiers: LCCN 2015040815| ISBN 9789814730280 (hardcover : alk. paper) |
ISBN 9789814730297 (pbk. : alk. paper)
Subjects: LCSH: Algebra, Abstract--Textbooks. | Algebra--Textbooks.
Classification: LCC QA162 .U53 2016 | DDC 512--dc23
LC record available at http://lccn.loc.gov/2015040815

British Library Cataloguing-in-Publication Data
A catalogue record for this book is available from the British Library.

In-house Editor: Bai Li

Printed in Singapore

to my son, Andre

Preface

The purpose of this book is to provide a concise yet detailed account of fundamental concepts in modern algebra. The target audience for this book is first-year graduate students in mathematics, though the first two chapters are probably accessible to well-prepared undergraduates.

The book contains five chapters. In Chapter 1 we cover groups, subgroups, quotient groups, homomorphisms of groups, and group structure, including cyclic groups, the Structure Theorem for finitely generated Abelian groups, Cauchy's Theorem, and Sylow's Theorems. In Chapter 2 we consider rings, the group of units of a ring, ideals, quotient rings, and ring homomorphisms. Included also are sections on localizations and completions. In Chapter 3 we turn to modules. We begin with a review of both finite and infinite dimensional vector spaces, and then generalize to modules over PIDs and Noetherian rings. We include sections on projective modules, tensor products of modules, algebras, and the discriminant of modules over an integral domain. In Chapter 4 we define simple algebraic extensions of $\mathbb{Q}$ and introduce the Galois group of the splitting field of a monic irreducible polynomial over $\mathbb{Q}$. We state and prove the Fundamental Theorem of Galois Theory. We then follow with an introduction (essentially) to algebraic number theory: we include material on the ring of integers of an algebraic extension, the Noetherian propery of the ring of integers, Dedekind domains and unique factorization of ideals. In the final chapter (Chapter 5) we cover the basic theory of finite fields and linearly recursive sequences.

We begin each chapter with an overview of the material to be covered. At the end of each chapter we give an extensive list of exercises which range from basic applications of the theory, to problems designed to challenge the reader. We also include some "Questions for Further Study", which are advanced problems suitable for master's level research projects.

I would like to thank the fellow algebraists who read and commented on earlier drafts of the manuscript. Their suggestions, especially those regarding the organization of the sections, have been duly noted and incorporated into the book. My appreciation is also extended to E. H. Chionh and Li Bai, at World Scientific, who have skillfully guided me through the publication process. To my wife, Rebecca Brower, who is also an academic, and who certainly understands the challenge of a writing project of this sort, I thank you for your patience, kindness and companionship. Any finally, to my son Andre, to whom this book is dedicated, I thank you for understanding that although writing takes a lot of time, in the end it is a worthy endeavor.

Robert G. Underwood

Contents

Chapter 1

Groups

In this chapter we introduce semigroups, monoids, and groups, give some basic examples of groups and discuss some of their elementary properties. We then consider subgroups, cosets and Lagrange's theorem, normal subgroups and the quotient group. We next turn to the basic maps between groups: homomorphisms and isomorphisms and their kernels. (Throughout this book, map = function.) We give the First, Second and Third Isomorphism theorems and the Universal Mapping Property for Kernels.

We close the chapter with the study of group structure, including generating sets for groups and subgroups and the notion of a cyclic group. From the cyclicity of the additive group of integers Z we obtain greatest common divisors, least common multiples, Bezout's Lemma and the Chinese Remainder Theorem. We state the structure theorem for finitely generated abelian groups. Regarding the structure of groups in general, we introduce G-sets, and give Cauchy's Theorem and Sylow's First, Second, and Third Theorems.

1.1 Introduction to Groups

In this section we define semigroups and monoids and give some examples, including the monoid of words on a finite alphabet. From semigroups and monoids, we develop the concept of a group, discuss finite, infinite and abelian groups, and prove some elementary properties of groups. We introduce examples of groups that we will appear throughout this book, including the additive group of integers, Z, the multiplicative group of non-zero real numbers, $\mathbb{R}^\times$ and the group of residue classes modulo n, Z_n. For further examples of groups we construct the 3rd and 4th dihedral groups, D_3, D_4 as the groups of symmetries of the equilateral triangle and the square, as well as the symmetric group on n letters, S_n.

* * *

Let S be a non-empty set of elements. The cartesian product on S is defined as $S \times S = \{(a, b) : a, b \in S\}$.

Definition 1.1. A **binary operation on** S is a function $B : S \times S \to S$; we denote the image of (a, b) by ab.

A binary operation is **commutative** if for all $a, b \in S$, $ab = ba$. A binary operation is **associative** if for all $a, b, c \in S$, $a(bc) = (ab)c$.

Definition 1.2. A **semigroup** is a set S together with an associative binary operation $S \times S \to S$.

Let S be a semigroup and let $a_1, a_2, a_3 \in S$. We define the product $a_1a_2a_3$ to be the common value of the expressions $(a_1a_2)a_3$ and $a_1(a_2a_3)$. For $n \geq 4$ we define the **product of elements** $a_1, a_2, \ldots, a_n \in S$ inductively to be

$$\prod_{i=1}^{n} a_i = \left(\prod_{i=1}^{n-1} a_i\right) a_n.$$

In defining $\prod_{i=1}^{n} a_i$ in this way we are asserting that we can insert parentheses into the product in any manner we choose without changing its value. For example, $a_1a_2a_3a_4$ is the common value of the expressions

$$(a_1a_2a_3)a_4,\ (a_1(a_2a_3))a_4,\ ((a_1a_2)a_3)a_4,\ a_1((a_2a_3)a_4),\ (a_1a_2)(a_3a_4),$$

$$a_1(a_2(a_3a_4)),\ a_1(a_2a_3)a_4,\ (a_1a_2)a_3a_4,\ a_1a_2(a_3a_4),\ a_1(a_2a_3a_4).$$

Definition 1.3. A **monoid** is a semigroup S in which there exists an element $e \in S$ with $ea = a = ae, \forall a \in S$. Such an element e is called an **identity element** for the monoid.

For example, the set of integers Z together with ordinary multiplication is a monoid with identity element $e = 1$ and the set of natural numbers $\mathbb{N} = \{1, 2, 3, \ldots\}$ together with ordinary addition is a semigroup. Note that $\mathbb{N}$ together with $+$ is not a monoid, however.

Here is an example of a monoid that is used in computer science. An **alphabet** Σ_0 is a non-empty set whose elements are the **letters** of the alphabet. A **word** is a finite sequence of letters in Σ_0. For a given alphabet Σ_0, let Σ_0^* denote the collection of all words formed from the alphabet Σ_0.

For $w \in \Sigma_0^*$, the **length of** w denoted by $l(w)$ is the number of letters in w. The **empty word** e is the (unique) word of length 0 in Σ_0^*. We endow Σ_0^* with a binary operation $\Sigma_0^* \times \Sigma_0^* \to \Sigma_0^*$ called **concatenation**. Concatenation (sometimes denoted as '$\cdot$') is defined as $x \cdot y = xy$, for $x, y \in \Sigma_0^*$. As the reader can easily verify, Σ_0^* together with concatenation is a monoid; the identity element is the empty word.

For example if $\Sigma_0 = \{a, b\}$, then $\{a, b\}^*$ consists of all finite sequences of a's and b's. The word $x = abbab \in \{a, b\}^*$ has length $l(x) = 5$. Moreover, if $y = bab$, then $x \cdot y = abbab \cdot bab = abbabbab$.

Definition 1.4. A **group** is a set G together with a binary operation $G \times G \to G$ for which

(i) the binary operation is associative,

(ii) there exists an element $e \in G$ for which $ea = a = ae$, for all $a \in G$,

(iii) for each $a \in G$, there exists an element $c \in G$ for which $ca = e = ac$.

An element e satisfying (ii) is an **identity element** for G; an element c satisfying (iii) is called an **inverse element** of a and is denoted by a^{-1}.

We note immediately that every group is a monoid. The converse is false, of course (see §1.6, Exercise 6).

There are many familiar examples of groups encountered in mathematics. For example, the set of integers Z, together with ordinary addition $+$ is a group, 0 plays the role of e, and $-a$ is the inverse of $a \in Z$. One easily shows that the set of rational numbers $\mathbb{Q}$ under ordinary addition and the set of real numbers $\mathbb{R}$ under ordinary addition are groups. The set of non-zero real numbers $\mathbb{R}^\times$ is a group under ordinary multiplication $\cdot$ with $e = 1$, and $a^{-1} = 1/a$. A further example is the **general linear group** $GL_n(\mathbb{R})$ consisting of invertible $n \times n$ matrices with entries in $\mathbb{R}$, together with matrix multiplication. Recalling some linear algebra, one has

$$GL_n(\mathbb{R}) = \{A \in Mat_n(\mathbb{R}) : \det(A) \neq 0\}.$$

In the case that $n = 1$, $GL_1(\mathbb{R}) = \mathbb{R}^\times$.

The **order** of a group G, denoted by $|G|$, is the number of elements in G. If $|G|$ is infinite, then G is an **infinite group**. All of the examples of groups given above are infinite groups. A group G is **finite** if $|G|$ is finite. In what follows we give an example of a finite group.

Let n, a be integers with $n > 0$. A **residue of** a **modulo** n is an integer r for which $a = nq + r$ for some $q \in Z$. For instance, if $n = 3$, $a = 8$, then 11 is a residue of 8 modulo 3 since $8 = 3(-1) + 11$, but so is 2 since $8 = 3(2) + 2$. The possible least non-negative residues of a modulo n are $0, 1, 2, \ldots, n-1$. The least non-negative residue of a modulo n is denoted as a **mod** n. For example, $8 \bmod 3 = 2$, but also note that $-3 \bmod 4 = 1$ and $11 \bmod 4 = 3 \bmod 4 = 3$. We say that two integers a, b are **congruent modulo** n if $a \bmod n = b \bmod n$ and we write $a \equiv b \bmod n$. Let a, n be integers with $n > 0$. Then n **divides** a, denoted by $n \mid a$, if there exists an integer k for which $a = nk$.

Proposition 1.1. *Let $a, b, n \in Z$, $n > 0$. Then $a \equiv b \bmod n$ if and only if $n \mid (a - b)$.*

Proof. To prove the "only if" part, assume that $a \equiv b \bmod n$. Then $a \bmod n = b \bmod n$, so there exist integers l, m for which $a = nm + r$ and $b = nl + r$ with $r = a \bmod n = b \bmod n$. Thus $a - b = n(m - l)$. For the "if" part, assume that $a - b = nk$ for some k. Then $(nm + a \bmod n) - (nl + b \bmod n) = nk$ for some $m, l \in Z$, so that n divides $a \bmod n - b \bmod n$. Consequently, $a \bmod n - b \bmod n = 0$, hence $a \equiv b \bmod n$. □

Proposition 1.1 can help us compute $a \bmod n$. For instance $-14 \bmod 17 = 3 \bmod 17 = 3$ since $17 \mid (-14 - 3)$. Likewise $-226 \bmod 17 = 12 \bmod 17 = 12$ since $17 \mid (-226 - 12)$.

For $n > 0$ consider the set $J = \{0, 1, 2, 3, \ldots, n-1\}$ of least non-negative residues modulo n. Note that $a = a \bmod n, \forall a \in J$. On J we define a binary operation $+_n$ as follows: for $a, b \in J$,

$$a \bmod n +_n b \bmod n = (a + b) \bmod n.$$

Then $+_n$ gives J the structure of a group, known as the **group of residue classes modulo** n. We denote this group by Z_n; Z_n is a finite group of order $|Z_n| = n$. For example, $Z_4 = \{0, 1, 2, 3\}$ and one has $1 +_4 2 = 3$, $3 +_4 2 = 1$, and so on.

One nice feature of a small finite group is that all possible group products can be arranged in a finite table in which the elements of the group are listed across the top as labels of the columns and down the left side as labels of the rows. For elements a, b in finite group G, the (a, b)th entry in the table is ab. This table is the **group table** for finite group G. For instance, the group table for Z_4 is

$+_4$	0	1	2	3
0	0	1	2	3
1	1	2	3	0
2	2	3	0	1
3	3	0	1	2

We can construct a new group from a finite set of groups. Let $S_1, S_2, \ldots, S_k$ be a finite collection of sets. Then the **cartesian product** $\prod_{i=1}^{k} S_i$ is the collection of all k-tuples $\{(a_1, a_2, \ldots, a_k) : a_i \in S_i\}$.

Proposition 1.2. *Let G_i, $i = 1, \ldots, k$, be a finite collection of groups. Then the cartesian product $\prod_{i=1}^{k} G_i$ is a group under the binary operation defined as*

$$(a_1, a_2, \ldots, a_k) \cdot (b_1, b_2, \ldots, b_k) = (a_1 b_1, a_2 b_2, \ldots, a_k b_k),$$

where $a_i b_i$ is the image of (a_i, b_i) under the binary operation $B_i : G_i \times G_i \to G_i$ of the group G_i, $1 \leq i \leq k$.

Proof. We show that the conditions of Definition 1.4 hold. Clearly the binary operation on the cartesian product is associative; for an identity element we take $e = (e_1, e_2, \ldots, e_k)$ where e_i is an identity in G_i. Lastly, for each k-tuple $(a_1, a_2, \ldots, a_k)$ one has $(a_1, a_2, \ldots, a_k)^{-1} = (a_1^{-1}, a_2^{-1}, \ldots, a_k^{-1})$. □

The group $\prod_{i=1}^{k} G_i$ of Proposition 1.2 is the **direct product group**.

As an illustration we consider the group $Z \times Z$ in which the binary operation is given as $(m_1, m_2) + (n_1, n_2) = (m_1 + n_1, m_2 + n_2)$. For another example, we take $Z_2 \times Z_3$; here for instance, $(0, 1) + (1, 2) = (1, 0)$. Note that $|Z_2 \times Z_3| = 6$.

In any group the identity and the inverse of an element are unique.

Proposition 1.3. *Let G be a group. Then there exists a unique element e for which $ea = a = ae$, and for each $a \in G$, there exists a unique element a^{-1} for which $a^{-1} a = e = a a^{-1}$.*

Proof. Suppose there are two identities e_1 and e_2. Then with e_1 acting on the left, $e_1e_2 = e_2$. Also, with e_2 acting on the right, $e_1e_2 = e_1$. Thus $e_1 = e_2$.

Now suppose there exist two inverses a_1^{-1} and a_2^{-1} for a given element $a \in G$. Then $a_1^{-1}a = e = a_2^{-1}a$. Now multiplying on the right by a_1^{-1} yields $a_1^{-1} = a_2^{-1}$. □

Since $(ab)(b^{-1}a^{-1}) = e = (ab)(ab)^{-1}$, uniqueness of the inverse yields the **rule for inverses of products** in a group, that is: $(ab)^{-1} = b^{-1}a^{-1}$.

In a group the binary operation is by definition associative. It may or may not be commutative.

Definition 1.5. A group for which the binary operation is commutative is an **abelian** group.

For example, the residue class group Z_n is an abelian group, as are Z, $\mathbb{Q}$, and $\mathbb{R}$.

The easiest example of a non-abelian group is $GL_2(\mathbb{R})$. In this group, for example, we have

$$\begin{pmatrix} 1 & 1 \\ 0 & 1 \end{pmatrix} \begin{pmatrix} 0 & 1 \\ 1 & 0 \end{pmatrix} \neq \begin{pmatrix} 0 & 1 \\ 1 & 0 \end{pmatrix} \begin{pmatrix} 1 & 1 \\ 0 & 1 \end{pmatrix}.$$

For a finite non-abelian group, we consider the 3rd order **dihedral group**, which is denoted by D_3. The elements of D_3 are the six "symmetries" of the equilateral triangle $\triangle ABC$ (Figure 1.1) and consist of three clockwise rotations of 0°, 120°, and 240° about the center O of the triangle, represented by the elements ρ_0, ρ_1, ρ_2, together with three reflections through the perpendicular lines ℓ_1, ℓ_2, ℓ_3, represented by the elements μ_1, μ_2, μ_3, respectively. It is critical to realize that the rotations move the vertices of the triangle, yet the perpendicular lines remain fixed and do not move with the rotation of the triangle.

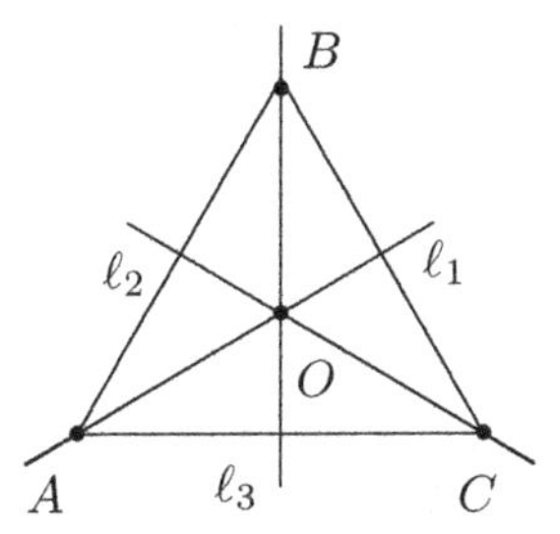

Fig. 1.1 Equilateral $\triangle ABC$, $\ell_1 \perp \overleftrightarrow{BC}$, $\ell_2 \perp \overleftrightarrow{AB}$, $\ell_3 \perp \overleftrightarrow{AC}$.

A rotation of 120°, followed by a reflection through the line ℓ_1 is equivalent to a reflection through the line ℓ_3, in other words, $\rho_1\mu_1 = \mu_3$ (Figure 1.2).

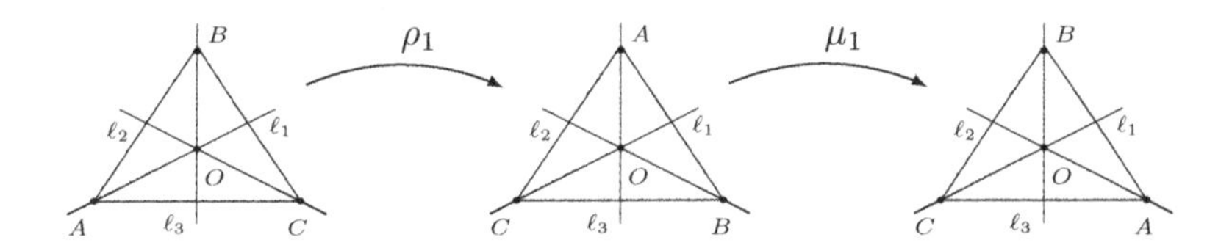

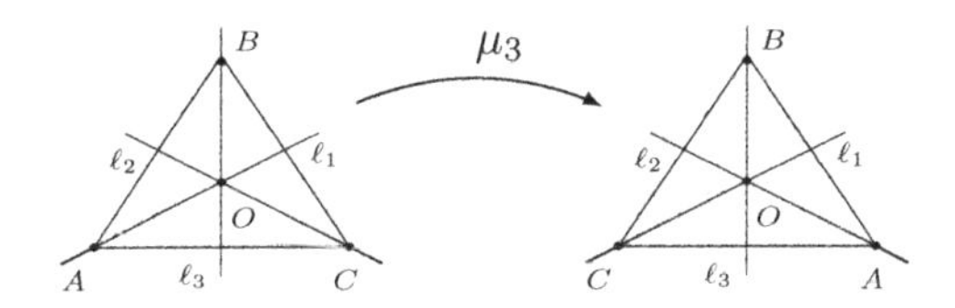

Fig. 1.2 In D_3, $\rho_1\mu_1 = \mu_3$.

The group table for D_3 is

	ρ_0	ρ_1	ρ_2	μ_1	μ_2	μ_3
ρ_0	ρ_0	ρ_1	ρ_2	μ_1	μ_2	μ_3
ρ_1	ρ_1	ρ_2	ρ_0	μ_3	μ_1	μ_2
ρ_2	ρ_2	ρ_0	ρ_1	μ_2	μ_3	μ_1
μ_1	μ_1	μ_2	μ_3	ρ_0	ρ_1	ρ_2
μ_2	μ_2	μ_3	μ_1	ρ_2	ρ_0	ρ_1
μ_3	μ_3	μ_1	μ_2	ρ_1	ρ_2	ρ_0

From the table we see that $\rho_1\mu_1 = \mu_3 \neq \mu_2 = \mu_1\rho_1$, and so D_3 is not abelian.

The 4th order dihedral group, denoted as D_4, consists of the eight symmetries of the square $\square ABCD$ (Figure 1.3). Elements of D_4 consist of four clockwise rotations of $0°$, $90°$, $180°$, and $270°$ about the center O of the square, represented by the elements ρ_0, ρ_1, ρ_2, and ρ_4, together with two reflections through the diagonal lines ℓ_1, ℓ_2, represented by the elements μ_1, μ_2, and two reflections through the perpendicular lines ℓ_3, ℓ_4, represented by the elements σ_1, σ_2, respectively. Again, it is critical to realize that the rotations move the vertices of the square, yet the lines of reflection remain fixed: they do not move with the rotation of the square.

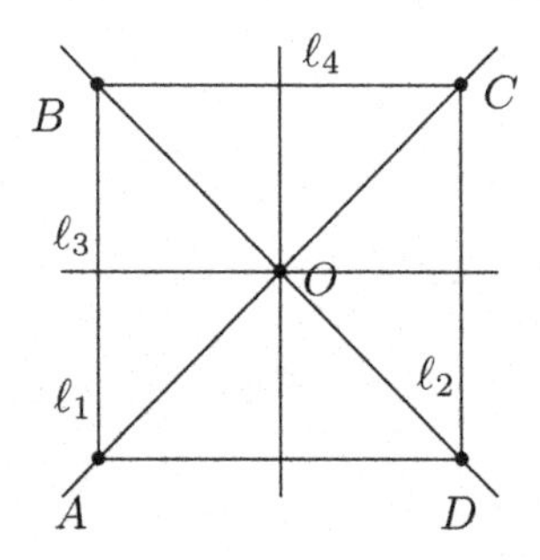

Fig. 1.3 Square $\square ABCD$, $\ell_1 = \overleftrightarrow{AC}$, $\ell_2 = \overleftrightarrow{BD}$, $\ell_3 \perp \overleftrightarrow{CD}$, $\ell_4 \perp \overleftrightarrow{AD}$.

In D_4, for example, we compute

$$\rho_3\sigma_2 = \mu_1 \neq \mu_2 = \sigma_2\rho_3.$$

Note that the result of applying an element of D_4 to the square preserves the square, yet mixes or permutes its vertices. For instance, the reflection σ_1 maps the vertices A, B, C, D to the vertices B, A, D, C. Here is a more general construction of an non-abelian group that mixes the elements of a set.

Let $\Sigma_0 = \{1, 2, 3, \ldots, n\}$ denote the set of n letters (actually they are integers, but we call them letters in this context). A **permutation** is a function $\sigma : \Sigma_0 \to \Sigma_0$ that is both one-to-one and onto (a bijection). Let S_n denote the set of all permutations on the set of n letters Σ_0. On S_n we define a binary operation $\circ$ which is ordinary function composition: for

$\sigma, \tau \in S_n$, $i \in \Sigma_0$,

$$(\sigma \circ \tau)(i) = \sigma(\tau(i)).$$

Proposition 1.4. *Let Σ_0 denote the set of n letters and let S_n denote the set of all permutations of Σ_0. Then S_n together with the binary operation $\circ$ is a group.*

Proof. We show that conditions of Definition 1.4 hold. For (i): Let $\sigma, \tau, \rho \in S_n$. Then

$$\sigma \circ (\tau \circ \rho) = (\sigma \circ \tau) \circ \rho,$$

since function composition is associative. For (ii), one takes an identity element e to be the identity permutation $\iota : \Sigma_0 \to \Sigma_0$, $x \mapsto x, \forall x \in \Sigma_0$. Then $\iota \circ \sigma = \sigma = \sigma \circ \iota$, as required. For (iii), since $\sigma : \Sigma_0 \to \Sigma_0$ is a bijection, the inverse map σ^{-1} exists and satisfies the property $\sigma^{-1} \circ \sigma = \iota = \sigma \circ \sigma^{-1}$. □

The group S_n given in Proposition 1.4 is the **symmetric group on n letters**. Observe that there are $n!$ possible permutations of the n letter set Σ_0 and so $|S_n| = n!$. An element $\sigma \in S_n$ can be written in **permutation notation** as

$$\sigma = \begin{pmatrix} 1 & 2 & 3 & 4 & \cdots & n-1 & n \\ \sigma(1) & \sigma(2) & \sigma(3) & \sigma(4) & \cdots & \sigma(n-1) & \sigma(n) \end{pmatrix}.$$

When elements of S_n are written in permutation notation, one can use "right-to-left" permutation multiplication to easily compute $\sigma \circ \tau$. For example, in S_5, let

$$\sigma = \begin{pmatrix} 1\,2\,3\,4\,5 \\ 2\,1\,4\,5\,3 \end{pmatrix}, \quad \tau = \begin{pmatrix} 1\,2\,3\,4\,5 \\ 4\,3\,2\,1\,5 \end{pmatrix}.$$

Then the right-to-left permutation multiplication is

$$\overleftarrow{\hphantom{xxxxxxxxxxxxxxxx}}$$

$$\sigma \circ \tau = \begin{pmatrix} 1\,2\,3\,4\,5 \\ 2\,1\,4\,5\,3 \end{pmatrix} \begin{pmatrix} 1\,2\,3\,4\,5 \\ 4\,3\,2\,1\,5 \end{pmatrix} = \begin{pmatrix} 1\,2\,3\,4\,5 \\ 5\,4\,1\,2\,3 \end{pmatrix},$$

since (reading right-to-left) $5 \mapsto 5 \mapsto 3$, $4 \mapsto 1 \mapsto 2$, and so on.

Proposition 1.5. *The symmetric group on n letters S_n is abelian for $n = 1, 2$ and non-abelian for $n \geq 3$.*

Proof. Exercise. □

For the permutation $\sigma \in S_5$ given above, one has

$$1 \overset{\sigma}{\mapsto} 2 \overset{\sigma}{\mapsto} 1,$$

and

$$3 \overset{\sigma}{\mapsto} 4 \overset{\sigma}{\mapsto} 5 \overset{\sigma}{\mapsto} 3.$$

Thus σ factors as

$$\sigma = (1,2)(3,4,5),$$

where $(1,2)$ and $(3,4,5)$ are shorthand for permutations:

$$(1,2) = \begin{pmatrix} 1\,2\,3\,4\,5 \\ 2\,1\,3\,4\,5 \end{pmatrix}, \quad \text{and} \quad (3,4,5) = \begin{pmatrix} 1\,2\,3\,4\,5 \\ 1\,2\,4\,5\,3 \end{pmatrix}.$$

Moreover, for $\tau \in S_5$,

$$1 \overset{\tau}{\mapsto} 4 \overset{\tau}{\mapsto} 1,$$

$$2 \overset{\tau}{\mapsto} 3 \overset{\tau}{\mapsto} 2,$$

and

$$5 \overset{\tau}{\mapsto} 5,$$

thus

$$\tau = (1,4)(2,3)(5),$$

where

$$(1,4) = \begin{pmatrix} 1\,2\,3\,4\,5 \\ 4\,2\,3\,1\,5 \end{pmatrix}, \quad (2,3) = \begin{pmatrix} 1\,2\,3\,4\,5 \\ 1\,3\,2\,4\,5 \end{pmatrix},$$

and

$$(5) = \begin{pmatrix} 1\,2\,3\,4\,5 \\ 1\,2\,3\,4\,5 \end{pmatrix}.$$

Permutations of the form $(1,2)$, $(3,4,5)$, $(1,4)$, and so on, are called **cycles**. Generally, a cycle in S_n is written $(a_1, a_2, \ldots, a_l)$ for distinct letters $a_i \in \Sigma_0$, and denotes the permutation

$$a_1 \mapsto a_2 \mapsto \cdots \mapsto a_{l-1} \mapsto a_l \mapsto a_1,$$

which fixes all other letters in S_n. The **length** of the cycle $(a_1, a_2, \ldots, a_l)$ is l. A cycle of length 2 is a **transposition**.

Every permutation in S_n can be written as a product of cycles, and ultimately, for $n \geq 2$, as a product of transpositions using the cycle decomposition formula

$$(a_1, a_2, \ldots, a_l) = (a_1, a_l)(a_1, a_{l-1}) \cdots (a_1, a_2).$$

For instance, $\sigma = (1,2)(3,5)(3,4)$. If $\sigma \in S_n$ can be written as a product of an even number of transpositions, then it is an **even permutation**, if σ factors into an odd number of transpositions, then σ is an **odd permutation**. Note that $\sigma \in S_5$ is odd, while τ is even.

1.2 Subgroups

In this section we consider subgroups, which are the analogs for groups of subsets of a set. We determine the collection of left and right cosets of a subgroup in a group, give the partition theorem and Lagrange's Theorem. We specialize to normal subgroups and construct the quotient group, whose elements are the left (or right) cosets of a subgroup.

* * *

Let H be a subset of a group G. Then the binary operation B on G restricts to a function $B|_H : H \times H \to G$. If $B|_H(H \times H) \subseteq H$, then H is **closed** under the binary operation B. In other words, H is closed under B if $B(a, b) = ab \in H$ for all $a, b \in H$. If H is closed under B, then $B|_H$ is a binary operation on H. Closure is fundamental to the next definition.

Definition 1.6. Let H be a subset of a group G that satisfies the following conditions.

(i) H is closed under the binary operation of G,

(ii) $e \in H$,

(iii) for all $a \in H$, $a^{-1} \in H$.

Then H is a **subgroup** of G, which we denote by $H \leq G$.

For example, $2Z = \{2n : n \in Z\} \leq Z$. The subset $\{0, 3, 6, 9\}$ is a subgroup of Z_{12}. The subset $\{\rho_0, \mu_1\}$ is a subgroup of D_3. The set of integers Z is a subgroup of the additive group $\mathbb{R}$.

Every group G admits at least two subgroups: the **trivial subgroup** $\{e\} \leq G$, and the group G which is a subgroup of itself. If $H \leq G$ and H is a proper subset of G, then H is a **proper subgroup** of G and we write $H < G$. Observe that the notation $H \leq G$ implies that H is a group under the restricted binary operation of G.

Definition 1.7. Let H be a subgroup of G with $a \in G$. The set of group products $aH = \{ah : h \in H\}$ is the **left coset of H in G represented by a.** The collection $Ha = \{ha : h \in H\}$ is the **right coset of H in G represented by a.**

Let aH be a left coset. The element $x \in G$ is a **representative of aH** if $xH = aH$.

Since $eH = He = H$, the subgroup H is always a left and right coset of itself in G represented by e. Observe that $1 + \{0,3,6,9\} = \{1,4,7,10\}$ and $2 + \{0,3,6,9\} = \{2,5,7,11\}$ are left cosets of $\{0,3,6,9\}$ in Z_{12} represented by 1 and 2, respectively. Also, $\{\rho_0, \mu_1\}$ and $\{\rho_0, \mu_1\}\mu_3 = \{\mu_3, \rho_2\}$ are right cosets of $\{\rho_0, \mu_1\}$ in D_3 represented by ρ_0 and μ_3, respectively. Considering Z as a subgroup of $\mathbb{R}$, one has the left coset $r + Z$, for $r \in \mathbb{R}$, as illustrated in Figure 1.4.

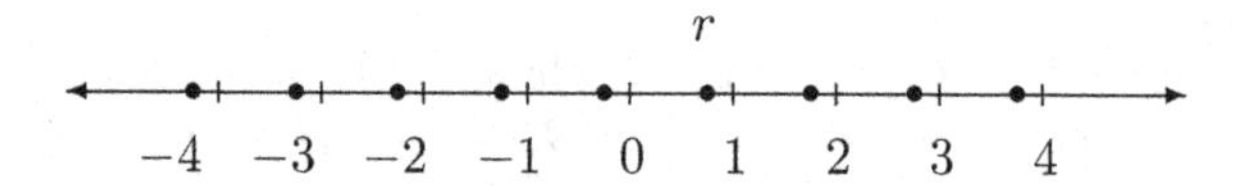

Fig. 1.4 The additive left coset $r + Z$ of Z in $\mathbb{R}$ represented by $r \in \mathbb{R}$.

Note that each real number r can be decomposed as $r = x + n$ for some $x \in [0,1)$ and some $n \in Z$. Thus $[0,1)$ is a complete set of representatives for the left cosets of Z in $\mathbb{R}$.

Proposition 1.6. *Let $H \leq G$, and let aH, bH be left cosets. Then there exists a bijection $\phi : aH \to bH$ defined as $\phi(ah) = bh$ for $h \in H$.*

Proof. To show that ϕ is 1-1 (one-to-one), suppose that $\phi(ah_1) = \phi(ah_2)$, for $h_1, h_2 \in H$. Then $bh_1 = bh_2$, so that $h_1 = h_2$, and consequently, $ah_1 = ah_2$. Next let $bh \in bH$. Then clearly, $\phi(ah) = bh$, so that ϕ is onto. □

The following corollary is immediate.

Corollary 1.1. *Suppose $|G| < \infty$. Then $|aH| = |bH| = |H|$ for all $a, b \in G$.*

Let G be a group, let H be a subgroup of G and let I be an arbitrary set. A subset $S = \{a_\eta\}_{\eta \in I}$ of G is a **left transversal** of H if the family $\{a_\eta H\}_{\eta \in I}$ constitutes the collection of all distinct left cosets of H in G.

Proposition 1.7. *Let G be a group, let H be a subgroup of G, and let $S = \{a_\eta\}_{\eta \in I}$ be a transversal of H. Then the collection of distinct left cosets of H in G forms a partition of the set G, that is,*

$$G = \bigcup_{\eta \in I} a_\eta H,$$

with $a_\eta H \cap a_\gamma H = \emptyset$ *whenever* $a_\eta H \neq a_\gamma H$.

Proof. Let $g \in G$. Then $gH = a_\eta H$ for some $\eta \in I$, thus $G \subseteq \bigcup_{\eta \in I} a_\eta H$. Clearly, $\bigcup_{\eta \in I} a_\eta H \subseteq G$, and so, $G = \bigcup_{\eta \in I} a_\eta H$. Suppose there exists an element $x \in a_\eta H \cap a_\gamma H$ for $\eta, \gamma \in I$. Then $x = a_\eta h_1 = a_\gamma h_2$ for some $h_1, h_2 \in H$. Consequently, $a_\eta = a_\gamma h_2 h_1^{-1} \in a_\gamma H$. Now, for any $h \in H$, $a_\eta h = a_\gamma h_1 h_2^{-1} h \in a_\gamma H$, and so, $a_\eta H \subseteq a_\gamma H$. By a similar argument $a_\gamma H \subseteq a_\eta H$, and so, $a_\eta H = a_\gamma H$. Thus the collection $\{a_\eta H\}_{\eta \in I}$ is a partition of G. □

To illustrate Proposition 1.7, let $G = D_3$, $H = \{\rho_0, \mu_1\}$. Then $H = \{\rho_0, \mu_1\}$, $\rho_1 H = \{\rho_1, \mu_3\}$ and $\rho_2 H = \{\rho_2, \mu_2\}$ are the distinct left cosets of H which form the partition of D_3,

$$\{H, \rho_1 H, \rho_2 H\}.$$

For another example, let $G = Z$, $H = 3Z$. Then the collection of distinct left cosets is $\{3Z, 1 + 3Z, 2 + 3Z\}$ which forms a partition of Z.

In many cases, even if the group G is infinite, there may be only a finite number of left cosets. When this occurs we define the number of left cosets of H in G to be the **index** $[G : H]$ **of** H **in** G. For instance, $[Z : 3Z] = 3$.

Proposition 1.8. *Let* $K \leq H \leq G$ *with* $[G : H] < \infty$, $[H : K] < \infty$. *Then* $[G : K] = [G : H][H : K]$.

Proof. Let $[G : H] = m$, $[H : K] = n$. By Proposition 1.7, there are partitions

$$G = a_1 H \cup a_2 H \cup \cdots \cup a_m H,$$

and

$$H = b_1 K \cup b_2 K \cup \cdots \cup b_n K.$$

Thus

$$G = \left(a_1 \bigcup_{j=1}^{n} b_j K\right) \cup \left(a_2 \bigcup_{j=1}^{n} b_j K\right) \cup \cdots \cup \left(a_m \bigcup_{j=1}^{n} b_j K\right) = \bigcup_{i,j=1}^{m,n} a_i b_j K$$

with $a_i b_j K \cap a_r b_s K = \emptyset$ if and only if $i = r$ and $j = s$. It follows that $[G : K] = mn$. □

If the group G is finite, we have the following classical result attributed to Lagrange.

Proposition 1.9 (Lagrange's Theorem). *Suppose $H \leq G$ with $|G| < \infty$. Then $|H|$ divides $|G|$.*

Proof. By Corollary 1.1 any two left cosets have the same number of elements. Since the left cosets partition G, we have $|H|[G : H] = |G|$. □

We haven't said much about right cosets and their relationship to left cosets. The method of Proposition 1.7 applies to show that the collection of distinct right cosets of H in G forms a partition of G with H as one of the cells. We also have the following propositions.

Proposition 1.10. *Let aH, Hb be left and right cosets of H in G. There is a bijection $\phi : aH \to Hb$ defined as $\phi(ah) = hb$ for all $h \in H$.*

Proof. Exercise. □

Let L denote the collection of all left cosets of H in G, and let R denote the collection of right cosets of H in G.

Proposition 1.11. *There exists a bijection $\phi : L \to R$ defined as $\phi(aH) = Ha^{-1}$.*

Proof. We first show that ϕ is well-defined on left cosets. Suppose that x is a representative of aH, that is, suppose that $xH = aH$. Then $x = ah$, for some $h \in H$, so that $x^{-1} = h^{-1}a^{-1} \in Ha^{-1}$. Thus $\phi(xH) = Hx^{-1} = Ha^{-1}$, so that ϕ is well-defined (ϕ is a function on cosets).

Now suppose $\phi(aH) = \phi(bH)$. Then $Ha^{-1} = Hb^{-1}$, hence $a^{-1} = hb^{-1}$ for some $h \in H$. Thus $a = bh^{-1} \in bH$, so that $aH = bH$. Thus ϕ is injective. Let $Hb \in R$, then $\phi(b^{-1}H) = Hb$, so ϕ is surjective. □

If G is abelian, then it is easy to see that $aH = Ha$ for all $a \in G$. But if G is non-abelian, one could have $aH \neq Ha$ for some $H \leq G$, $a \in G$. For example, consider $\{\rho_0, \mu_1\} \leq D_3$. Then

$$\rho_1\{\rho_0, \mu_1\} = \{\rho_1, \mu_3\} \neq \{\rho_0, \mu_1\}\rho_1 = \{\rho_1, \mu_2\}.$$

Definition 1.8. A subgroup $H \leq G$ is **normal** if $aH = Ha$ for all $a \in G$. In this case, we write $H \triangleleft G$.

A group G always has at least two normal subgroups: $\{e\}$ and G. One has that $3Z \lhd Z$, and $\{\rho_0, \rho_1, \rho_2\} \lhd D_3$. But $\{\rho_0, \mu_1\}$ is not a normal subgroup of D_3.

If $H \lhd G$, then for $h \in H$, $a \in G$, there exists an element $h' \in H$ with $ha = ah'$. The next three propositions illustrate the importance of normal subgroups.

Proposition 1.12. *Let H, K be subgroups of G with K normal in G. Then*

$$HK = \{hk : h \in H, k \in K\}$$

is a subgroup of G.

Proof. We show that the conditions of Definition 1.6 are satisfied. Let $h_1k_1, h_2k_2 \in HK$. Since $K \lhd G$, there exists an element $k_3 \in K$ with $k_1h_2 = h_2k_3$. Thus $h_1k_1h_2k_2 = h_1h_2k_3k_2 \in HK$, and so, HK is closed under the binary operation of G. Moreover, $e = ee \in HK$. Finally, let $hk \in HK$. Then $(hk)^{-1} = k^{-1}h^{-1} = h^{-1}k'$ for some $k' \in K$, and so, HK is closed under inverses. □

Let $H \leq G$ and let G/H denote the collection of left cosets of H in G. Define a relation

$$B : G/H \times G/H \to G/H$$

by the rule $B(aH, bH) = abH$, for $aH, bH \in G/H$.

Proposition 1.13. *Suppose that $H \lhd G$. Then the relation B defined as above is a binary operation on G/H.*

Proof. We show that B is well-defined on pairs of left cosets, that is, we show that $B : G/H \times G/H \to G/H$ is a function. Suppose that x and y are representatives of the left cosets aH and bH, respectively. We show that $B(xH, yH) = abH$. Observe that $x = ah_1$, $y = bh_2$ for some $h_1, h_2 \in H$. Thus

$$B(xH, yH) = xyH = ah_1bh_2H.$$

Since $H \lhd G$, there exists $h \in H$ so that $ah_1bh_2H = abhh_2H$. Clearly, $abhh_2H = abH$ which completes the proof. □

Proposition 1.14. *Let $H \lhd G$, and let G/H denote the collection of left cosets of H in G. Then G/H is a group under the left coset multiplication defined in Proposition 1.13.*

Proof. We need to show that conditions (i), (ii), and (iii) of Definition 1.4 hold. For (i), let $aH, bH, cH \in G/H$. Then

$$\begin{aligned} aH(bHcH) &= aHbcH \\ &= a(bc)H \\ &= (ab)cH \\ &= abHcH \\ &= (aHbH)cH. \end{aligned}$$

We leave it as exercises to show that $eH = H$ serves as the identity element in G/H and that $a^{-1}H$ is the inverse of aH under coset multiplication. □

This group of cosets given in Proposition 1.14 is the **quotient group G/H of G by H**. For example, since $nZ \triangleleft Z$, one has the additive quotient group Z/nZ. In the case $n = 3$, the group table for $Z/3Z$ is

	$3Z$	$1+3Z$	$2+3Z$
$3Z$	$3Z$	$1+3Z$	$2+3Z$
$1+3Z$	$1+3Z$	$2+3Z$	$3Z$
$2+3Z$	$2+3Z$	$3Z$	$1+3Z$

Observe that the quotient group $Z/3Z$ is essentially the same as the residue class group Z_3 (compare their group tables). For a multiplicative example, note that $H = \{\rho_0, \rho_1, \rho_2\} \triangleleft D_3$, hence the quotient group D_3/H exists. The group table for D_3/H is

	H	$\mu_1 H$
H	H	$\mu_1 H$
$\mu_1 H$	$\mu_1 H$	H

1.3 Homomorphisms of Groups

If A and B are sets, then the functions $f : A \to B$ are the basic maps between A and B. In this section we introduce group homomorphisms: functions preserving group structure which are the basic maps between groups. We consider isomorphisms of groups, isomorphism classes of groups, the First, Second, and Third Isomorphism Theorems, as well as the Universal Mapping Property for Kernels.

* * *

Definition 1.9. Let G, G' be groups. A map $\phi : G \to G'$ is a **homomorphism of groups** if for all $a, b \in G$, $\phi(ab) = \phi(a)\phi(b)$.

In additive notation, the homomorphism condition is given as

$$\phi(a+b) = \phi(a) + \phi(b).$$

For example, the map $\phi : Z \to Z/nZ$ given by $\phi(a) = a + nZ$ is a homomorphism of groups since

$$\phi(a+b) = a + b + nZ = (a + nZ) + (b + nZ) = \phi(a) + \phi(b).$$

The map $\phi : GL_n(\mathbb{R}) \to \mathbb{R}^\times$ defined as $\phi(A) = \det(A)$ is a homomorphism of groups since by a familiar property of determinants, $\phi(AB) = \det(AB) = \det(A)\det(B) = \phi(A)\phi(B)$. Another example comes from elementary calculus: The map $\phi : \mathbb{R}^+ \to \mathbb{R}$, defined by $\phi(x) = \ln(x)$ is a homomorphism of groups; it is an example of a homomorphism of a multiplicative group into an additive group.

Given the homomorphism $\phi(x) = \ln(x)$, one has $\phi(1) = 0$ and $\phi(x^{-1}) = -\phi(x)$ for all $x \in \mathbb{R}^+$. This illustrates two properties shared by all group homomorphisms.

Proposition 1.15. *Let $\phi : G \to G'$ be a group homomorphism. Then*

(i) $\phi(e) = e'$ where e' denotes the identity element of G',

(ii) $(\phi(a))^{-1} = \phi(a^{-1})$, for all $a \in G$.

Proof. For (i): $\phi(a) = \phi(ae) = \phi(a)\phi(e)$, so that $\phi(e) = e'$. Moreover, $e' = \phi(e) = \phi(aa^{-1}) = \phi(a)\phi(a^{-1})$, so that $(\phi(a))^{-1} = \phi(a^{-1})$ which proves (ii). □

The homomorphic image of a group is a group.

Proposition 1.16. *Let $\phi : G \to G'$ be a homomorphism of groups. Then $\phi(G) \le G'$.*

Proof. Exercise. □

Let $\phi : G \to G'$ be a group homomorphism. We sometimes denote the image $\phi(G)$ of ϕ as $\text{im}(\phi)$.

Definition 1.10. Let $\phi : G \to G'$ be a group homomorphism. The **kernel of** ϕ, denoted by $\ker(\phi)$, is the subset of G defined as $\{g \in G : \phi(g) = e'\}$.

Proposition 1.17. *The kernel of the homomorphism $\phi : G \to G'$ is a normal subgroup of G.*

Proof. We first show that $H = \ker(\phi)$ is a subgroup of G. Let $a, b \in H$. Then $\phi(ab) = \phi(a)\phi(b) = e$, hence H is closed under the binary operation of G. Also, Proposition 1.15(i) and (ii) show that $e \in H$ and that $a^{-1} \in H$ whenever $a \in H$, thus $H \leq G$.

Now suppose $ah \in aH$. Then $\phi(ah) = \phi(a)\phi(h) = \phi(a)e' = \phi(a)$. Thus aH consists of elements $x \in G$ with $\phi(x) = \phi(a)$. So for $x \in aH$, $e' = \phi(x)(\phi(a))^{-1} = \phi(x)\phi(a^{-1}) = \phi(xa^{-1})$, and hence, $xa^{-1} \in H$, which yields $x \in Ha$. So $aH \subseteq Ha$. By a similar argument, $Ha \subseteq aH$ which says that $H \triangleleft G$. □

The kernel of a homomorphism says much about the nature of the homomorphism.

Proposition 1.18. *Let $\phi : G \to G'$ be a homomorphism. Then ϕ is an injection if and only if* $\ker(\phi) = \{e\}$.

Proof. We prove the "if" part and leave the converse as an exercise. Suppose $\phi(x) = \phi(y)$ for $x, y \in G$. Then $\phi(x)(\phi(y))^{-1} = e'$, and so, $\phi(x)\phi(y^{-1}) = e'$ by Proposition 1.15(ii). Thus, $\phi(xy^{-1}) = e'$ which says that xy^{-1} is in the kernel of ϕ. Since $\ker(\phi)$ is trivial, $xy^{-1} = e$, that is, $x = y$. □

Definition 1.11. A homomorphism of groups $\phi : G \to G'$ which is both injective and surjective is an **isomorphism of groups.** Two groups G, G' are **isomorphic** if there exists an isomorphism $\phi : G \to G'$. We then write $G \cong G'$.

For example, $D_3 \cong S_3$ and Z_4 is isomorphic to the subgroup $\{\rho_0, \rho_1, \rho_2, \rho_3\} \leq D_4$ (see §1.6, Exercise 35).

Proposition 1.19. *Let $n > 0$. Then $\phi : Z/nZ \to Z_n$ defined by $a + nZ \mapsto a \bmod n$ is an isomorphism of groups.*

Proof. We first show that ϕ is well-defined, that is, we show that ϕ is a function. If $a + nZ = b + nZ$, then $n \mid (a - b)$ and so, by Proposition 1.1, $a \bmod n = b \bmod n$. Thus ϕ is well-defined.

Now,

$$\begin{aligned}\phi((a+nZ)+(b+nZ)) &= \phi(a+b+nZ)\\ &= (a+b) \bmod n\\ &= a \bmod n +_n b \bmod n\\ &= \phi(a+nZ) +_n \phi(b+nZ),\end{aligned}$$

so that ϕ is a homomorphism. Moreover, $\phi(a+nZ) = \phi(b+nZ)$ implies that $a \equiv b \bmod n$. Thus $a = b + nm$ for some $m \in Z$. It follows that $a + nZ = b + nZ$, so that ϕ is an injection. Clearly, ϕ is surjective. □

Proposition 1.20. *Let $n > 0$. Then $\phi : Z \to nZ$ with $\phi(a) = na$ is an isomorphism of additive groups.*

Proof. Note that

$$\phi(a+b) = n(a+b) = na + nb = \phi(a) + \phi(b),$$

$\forall a, b \in Z$, thus ϕ is a homomorphism. Now suppose $na = nb$. Then $na + (-nb) = n(a + (-b)) = 0$, hence $a = b$. Thus ϕ is an injection. Clearly ϕ is surjective. □

A map can of course be a bijection without being a group isomorphism. For example, there are exactly $6! = 120$ functions from D_3 to Z_6 that are bijections, yet none are group isomorphisms. Likewise, the groups Z_4 and $Z_2 \times Z_2$ have the same number of elements, but are not isomorphic.

Let G be any group. The collection of all groups G' for which $G' \cong G$ is the **isomorphism class of groups represented by** G. Essentially, two groups are contained in the same isomorphism class if and only if they are isomorphic. Our observation above shows that as groups D_3 and Z_6 are in different isomorphism classes.

There is an elegant relationship between group homomorphisms and quotient groups expressed in the following propositions.

Proposition 1.21. *Let $N \triangleleft G$. Then the map $\gamma : G \to G/N$ given by $\gamma(a) = aN$ is a surjective group homomorphism with kernel N.*

Proof. Let $aN \in G/N$. Since $\gamma(a) = aN$, γ is surjective. Now, $\gamma(ab) = abN = aNbN = \gamma(a)\gamma(b)$, which shows that γ is a homomorphism. Now by the definition of kernel,

$$\ker(\gamma) = \{a \in G : \gamma(a) = N\} = \{a \in G : aN = N\}.$$

Since $aN = N$ if and only if $a \in N$, $N = \ker(\gamma)$. □

Proposition 1.22 (First Isomorphism Theorem). *Let $\phi : G \to G'$ be a group homomorphism with $N = ker(\phi)$. Then $\gamma : G/N \to \phi(G)$ defined by $\gamma(aN) = \phi(a)$ is a group isomorphism.*

Proof. First note that G/N is a group since $N \lhd G$ (Proposition 1.14) and $\phi(G)$ is a group by Proposition 1.16. Suppose that $aN = bN$ for $a, b \in G$. Then $a = bn$ for some $n \in N$, and so, $\phi(a) = \phi(bn) = \phi(b)\phi(n) = \phi(b)$. Thus, γ is well-defined on cosets.

We have $\gamma(aNbN) = \gamma(abN) = \phi(ab) = \phi(a)\phi(b) = \gamma(aN)\gamma(bN)$, so γ is a homomorphism. It remains to show that γ is a bijection. Suppose $\gamma(aN) = \gamma(bN)$. Then $\phi(a) = \phi(b)$, hence $\phi(a)\phi(b)^{-1} = \phi(ab^{-1}) = e'$, which says that $ab^{-1} \in N$, and thus $a \in bN$. Hence $aN = bN$, so that γ is injective. Next let $b \in \phi(G)$. Then there exists an element $a \in G$ for which $\phi(a) = b$, thus $\gamma(aN) = b$, and γ is surjective. □

Proposition 1.23 (Universal Mapping Property for Kernels). *Let $\phi : G \to G'$ be a group homomorphism with $N = ker(\phi)$. Suppose that $K \leq N$ and $K \lhd G$. Then there exists a surjective homomorphism of groups $\psi : G/K \to \phi(G)$ defined by $\psi(aK) = \phi(a)$.*

Proof. We only need to check that ψ is well-defined. Suppose that $aK = bK$ for $a, b \in G$. Then $a = bk$ for some $k \in K$, and so, $\phi(a) = \phi(bk) = \phi(b)\phi(k) = \phi(b)$ since $k \in K \leq N$. □

Let $\phi : G \to G'$ be a group homomorphism with $N = \ker(\phi)$. The Universal Mapping Property for Kernels (UMPK) says that given a subgroup $K \leq N$, $K \lhd G$, there exists a group homomorphism $\psi : G/K \to G'$ so that

$$\psi s = \phi$$

where $s : G \to G/K$ is the canonical surjection; we say that ϕ "factors through" G/K and the following diagram commutes:

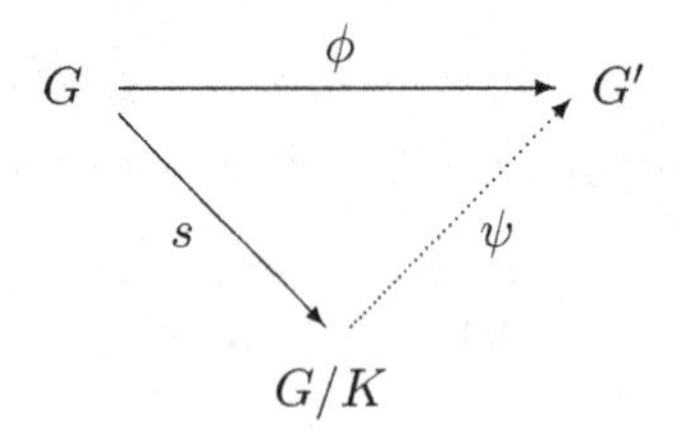

Here are two applications of the First Isomorphism Theorem.

Proposition 1.24 (Second Isomorphism Theorem). *Let H and K be subgroups of G and suppose that K is normal in G. Then*

(i) $K \triangleleft HK$,

(ii) $(H \cap K) \triangleleft H$,

(iii) $HK/K \cong H/(H \cap K)$.

Proof. To prove (i): By Proposition 1.12, $HK \leq G$, and so, $K \triangleleft HK$ since $K \triangleleft G$. For (ii): One easily obtains $(H \cap K) \leq G$ (we leave the details as an exercise). Let $ag \in a(H \cap K)$, $a \in H$. Then $ag = ha$ for some $h \in H$ since $H \triangleleft H$, and $ag = ka$ for some $k \in K$ since $K \triangleleft H$. Multiplying on the right by a^{-1} yields $h = k$, and so, $ag \in (H \cap K)a$. By a similar argument, $(H \cap K)a \subseteq a(H \cap K)$, and so $(H \cap K) \triangleleft H$.

We turn to statement (iii). The canonical surjection $\gamma : G \to G/K$ restricts to a homomorphism $\gamma|_H : H \to G/K$, whose kernel is precisely $H \cap K$. By Proposition 1.22, $H/(H \cap K) \cong \gamma(H)$. At the same time, γ restricts to a homomorphism $\gamma|_{HK} : HK \to G/K$, whose image is $\gamma(H)$. Since the kernel of $\gamma|_{HK}$ is K, $HK/K \cong \gamma(H)$, by Proposition 1.22. Thus, $HK/K \cong H/(H \cap K)$. □

Proposition 1.25 (Third Isomorphism Theorem). *Let $K \leq H \leq G$ and suppose that $K \triangleleft G$, and $H \triangleleft G$. Then $K \triangleleft H$, $(H/K) \triangleleft (G/K)$, and*

$$(G/K)/(H/K) \cong G/H.$$

Proof. Let $\gamma : G \to G/H$ be the canonical surjection given by Proposition 1.21. Note that $H = \ker(\gamma)$ and $\gamma(G) = G/H$. Now, $K \leq H$ and $K \triangleleft G$, and so, by the Universal Mapping Property for Kernels (UMPK), there is a surjective group homomorphism $\psi : G/K \to G/H$, defined as $\psi(aK) = aH$. Now, $\ker(\psi) = H/K$, and so by Proposition 1.22, $(G/K)/(H/K) \cong G/H$. Observe that this isomophism is given as $(aK)(H/K) \mapsto \psi(aK) = aH$. □

For example, Proposition 1.25 says that $(Z/6Z)/(3Z/6Z) \cong Z/3Z$ under the isomorphism $(n + 6Z) + 3Z/6Z \mapsto n + 3Z$.

A sequence of homomorphisms and groups of the form

$$1 \xrightarrow{\gamma_0} G_1 \xrightarrow{\gamma_1} G_2 \xrightarrow{\gamma_2} G_3 \xrightarrow{\gamma_3} 1$$

in which $\mathrm{im}(\gamma_i) = \ker(\gamma_{i+1})$ for $i = 0, 1, 2$ is a **short exact sequence of homomorphisms and groups**. As an example, we have the short exact sequence of additive abelian groups

$$0 \longrightarrow 2Z \longrightarrow Z \longrightarrow Z/2Z \longrightarrow 0,$$

where the maps are the obvious ones. For another example, consider the short exact sequence

$$0 \longrightarrow \{0,3\} \longrightarrow Z_6 \xrightarrow{s} Z_6/\{0,3\} \longrightarrow 0,$$

again with the obvious maps. This sequence is special, it is "split", that is, there exists an injective group homomorphism $l : Z_6/\{0,3\} \to Z_6$ for which the composition $s \circ l$ is the identity map on $Z_6/\{0,3\}$. Indeed, one may take $l : Z_6/\{0,3\} \to Z_6$ defined as $l(\{0,3\}) = 0$, $l(\{1,4\}) = 4$, $l(\{2,5\}) = 2$. Consequently, $Z_6 \cong \{0,3\} \times Z_6/\{0,3\}$. Since $Z_2 \cong \{0,3\}$ and $Z_3 \cong Z_6/\{0,3\}$, this becomes $Z_6 \cong Z_2 \times Z_3$, as we shall see in the next section.

1.4 Group Structure

In this section we discuss the structure of groups, define generating sets for groups and subgroups, and the concept of a cyclic group. The additive group Z is cyclic and from this we obtain greatest common divisors, least common multiples and Bezout's Lemma. Using the First Isomorphism Theorem we proof the Chinese Remainder Theorem, leading to the decomposition theorem for Z_n. We state the structure theorem for finitely generated abelian groups.

* * *

Let G be a group, and let $n > 0$ be an integer. We set $a^n = \underbrace{aa\cdots a}_{n}$, and $a^{-n} = \underbrace{a^{-1}a^{-1}\cdots a^{-1}}_{n}$. Since $e = a^{-1}a$, we put $a^0 = e$. In additive notation, one writes $na = \underbrace{a + a + \cdots + a}_{n}$, $-na = \underbrace{(-a) + (-a) + \cdots + (-a)}_{n}$, and $0a = e$.

Let S be a non-empty subset of G and let S^{-1} be the set of inverses of elements of S. Let $(S \cup S^{-1})^*$ denote the collection of all words of finite length built from the letters in $S \cup S^{-1}$. For instance, if $S = \{a, b\}$, then the elements

$$a^2, ab, ba^{-1}, a^{-3}, a^2b, ab^2, a^{-1}b^{-2}a^2, b^{-1}ab, b^3, ba^2$$

are in $(S \cup S^{-1})^*$. As in §1.1, $(S \cup S^{-1})^*$ is a monoid under concatenation which is now identified with the group product in G.

Proposition 1.26. *Let G be a group and let S be a non-empty subset of G. Then $(S \cup S^{-1})^*$ is a subgroup of G.*

Proof. We show that the conditions of Definition 1.6 hold. Let

$$x = a_1 a_2 \cdots a_k, \quad y = b_1 b_2, \ldots, b_l$$

be elements in $(S \cup S^{-1})^*$. Then

$$xy = a_1 a_2 \cdots a_k b_1 b_2, \ldots, b_l$$

is in $(S \cup S^{-1})^*$, and so $(S \cup S^{-1})^*$ is closed under the binary operation of G. For $a \in S$, $e = aa^{-1} \in (S \cup S^{-1})^*$. Lastly, by the rule for inverses of products,

$$\begin{aligned} x^{-1} &= (a_1 a_2 \cdots a_k)^{-1} \\ &= a_k^{-1} a_{k-1}^{-1} \cdots a_1^{-1}, \end{aligned}$$

which is an element of $(S \cup S^{-1})^*$. □

The subgroup $(S \cup S^{-1})^*$ of G is called the **subgroup of G generated by S** and is denoted by $\langle S \rangle$. For example, if $G = D_3$, $S = \{\rho_2, \mu_1\}$, then $\langle S \rangle = D_3$ since

$$\rho_0 = \rho_2^3, \ \rho_1 = \rho_2^2, \ \rho_2 = \rho_2, \ \mu_1 = \mu_1, \ \mu_2 = \rho_2\mu_1, \ \mu_3 = \mu_1\rho_2.$$

If $S = \{\rho_2\} \subseteq D_3$, then

$$\langle S \rangle = \{\rho_2^n : \ n \in Z\} = \{\rho_0, \rho_1, \rho_2\}.$$

If G is abelian, then $\langle S \rangle$ has a somewhat simpler definition. Let $S = \{a_\beta\}_{\beta \in I}$ be a non-empty subset of G indexed by the set I. Then $\langle S \rangle$ consists of all group products of the form

$$x = \prod_{\beta \in I} a_\beta^{n_\beta},$$

where the n_β are integers that are 0 for all but a finite number of subscripts β. If G is an additive group (and hence necessarily abelian), $\langle S \rangle$ has a form familiar to students of linear algebra: it is the collection of all quantities of the form

$$x = \sum_{\beta \in I} n_\beta a_\beta,$$

where $n_\beta = 0$ for all but a finite number of β, known as the **Z-linear combinations** of the set S.

Better yet, if G is abelian and $S = \{a_1, a_2, \ldots, a_k\}$ is finite, then $\langle S \rangle$ consists of all group products of the form

$$x = \prod_{i=1}^{k} a_i^{n_i},$$

for integers n_i. In additive notation this is:

$$x = \sum_{i=1}^{k} n_i a_i.$$

Let G be any group and let $S = \{a\}$ be a singleton subset of G. Then the subgroup $\langle S \rangle$ is **the cyclic subgroup of G generated by** a. Note that $\langle S \rangle = \langle \{a\} \rangle = \{a^n : n \in Z\}$. We usually denote $\langle \{a\} \rangle$ as $\langle a \rangle$. For instance, in D_3, $\langle \rho_1 \rangle = \{\rho_0, \rho_1, \rho_2\}$. In $\mathbb{R}^+$, under ordinary multiplication,

$$\langle 2 \rangle = \{2^n : n \in Z\} = \{\ldots, \frac{1}{8}, \frac{1}{4}, \frac{1}{2}, 1, 2, 4, 8, \ldots\} \leq \mathbb{R}^+.$$

For an additive example, take $G = Z$. Then

$$\langle -3 \rangle = \{n(-3) : n \in Z\} = \{\ldots, 9, 6, 3, 0, -3, -6, -9 \ldots\} \leq Z.$$

Let G be a group and let $a \in G$. The **order** of a is the order of the cyclic subgroup $\langle a \rangle \leq G$.

Proposition 1.27. *Suppose G is finite with $n = |G|$. Let $g \in G$. Then $g^n = e$.*

Proof. Let $H = \langle g \rangle$ denote the cyclic subgroup generated by g. Let $m = |\langle g \rangle|$ and let

$$1 = g^0, g^1, g^2, \ldots, g^{m-1} \tag{1.1}$$

be the list of powers of g. We claim that list (1.1) contains m distinct powers of g. If not, let s be the smallest integer $1 \leq s \leq m-1$ for which the list

$$1 = g^0, g^1, g^2, \ldots, g^{s-1} \tag{1.2}$$

contains distinct powers of g. Now, $g^s = g^i$ for some i, $0 \leq i \leq s-1$. Thus $g^{s-i} = 1$. Now, if $i \geq 1$, then $1 \leq s-i \leq s-1$ with $g^{s-i} = g^0 = 1$, which constradicts our assumption that list (1.2) contains distinct powers of g. Thus $i = 0$ and so, $g^s = 1$. It follows that the powers in (1.2) constitute the cyclic subgroup $\langle g\rangle$, but this says that $|\langle g\rangle| = s < m$, a contradiction. We conclude that the list (1.1) contains distinct powers of g. Now $g^m = g^i$ for some $0 \leq i \leq m-1$ and arguing as above, we conclude that $i = 0$. Thus $g^m = e$.

Now by Lagrange's Theorem $m \mid n$, that is, $ml = n$ for some integer l. Hence $g^n = g^{ml} = (g^m)^l = e$. □

Let $S \subseteq G$. Then G is **generated by** S if $G = \langle S\rangle$. Every group is generated by some subset of the group. If necessary, one could choose the set G as a generating set for itself.

Suppose G is generated by the set S. Then G is **finitely generated** if S is finite, that is, if $S = \{a_1, a_2, \ldots, a_k\}$, for some integer k. We then write $G = \langle a_1, a_2, \ldots, a_k\rangle$. For example, D_3 is finitely generated by $S = \{\rho_2, \mu_1\}$; we have $D_3 = \langle \rho_2, \mu_1\rangle$. If G is abelian and finitely generated by the set $S = \{a_1, a_2, \ldots, a_k\}$ then every element of G has the form

$$\prod_{i=1}^{k} a_i^{n_i} = a_1^{n_1} a_2^{n_2} \cdots a_k^{n_k}, \quad n_i \in Z.$$

If G is additive, then this becomes

$$\sum_{i=1}^{k} n_i a_i = n_1 a_1 + n_2 a_2 + \cdots n_k a_k, \quad n_k \in Z.$$

A group G is **cyclic** if it is generated by a singleton set $S = \{a\}$. In this case, $G = \langle a\rangle$, with $\langle a\rangle = \{a^n : n \in Z\}$. For example, Z is cyclic since $Z = \langle 1\rangle = \{n1 : n \in Z\}$ and Z_n is cyclic since $\langle 1\rangle = Z_n$.

Let $G = \langle a\rangle$, $G' = \langle a'\rangle$ be cyclic groups. Then $G \cong G'$, since $\phi : G \to G'$ defined by $a^i \mapsto (a')^i$ is an isomorphism.

Let n be a positive integer. The **cyclic group of order** n **on the generator** g is the group C_n consisting of the set of elements

$$C_n = \{1 = g^0, g = g^1, g^2, g^3, \ldots, g^{n-1}\},$$

together with the binary operation defined as

$$g^i g^j = g^{(i+j) \bmod n}.$$

Note that $g^n = 1$. For example, the group table for C_4 is

	1	g	g^2	g^3
1	1	g	g^2	g^3
g	g	g^2	g^3	1
g^2	g^2	g^3	1	g
g^3	g^3	1	g	g^2

Much is known about the structure of cyclic groups.

Proposition 1.28. *Every cyclic group is abelian.*

Proof. Exercise. □

Proposition 1.29. *Let p be a prime number and let G be a finite group of order p. Then G is cyclic.*

Proof. Let $a \in G$, $a \neq e$. Let $\langle a \rangle$ denote the cyclic subgroup of G generated by a. By Lagrange's Theorem $|\langle a \rangle|$ divides p. Since $|\langle a \rangle| > 1$, $\langle a \rangle = G$. Thus G is cyclic. □

A group of order 2, 3, or 5 is cyclic by Proposition 1.29, hence abelian by Proposition 1.28. Thus a group of order 2, 3, 5 is isomorphic to Z_2, Z_3, Z_5, respectively. What about groups of order 4?

Proposition 1.30. *Let G be a group of order* 4. *Then G is abelian.*

Proof. Let $a \neq e$ in G. If $a^2 \neq e$, then a has order 4 in G, hence G is cyclic and consequently abelian by Proposition 1.28. So assume that $a^2 = e$ and let b be some other element of G not equal to a or e. Consider ab. If $ab = e$ then $a = b$ by uniqueness of the inverse of a; if $ab = a$, then $b = e$ by left cancellation (multiply both sides on the left by a^{-1}). If $ab = b$, then $a = e$ be right cancellation. Thus ab is the fourth element of G. By a similar argument ba is the fourth element of G, thus $ab = ba$, and so G is abelian. □

Thus a group G of order 4 is abelian. If G is cyclic, then it is isomorphic to Z_4. If G is not cyclic, then all three non-trivial elements have order 2, and consequently, $G \cong Z_2 \times Z_2$. To represent the isomorphism class containing $Z_2 \times Z_2$ we sometimes choose the **Klein 4-group**, $\mathcal{V} = \{e, a, b, c\}$. The group table for $\mathcal{V}$ is

	e	a	b	c
e	e	a	b	c
a	a	e	c	b
b	b	c	e	a
c	c	b	a	e

As we have seen, the 3rd and 4th order dihedral groups D_3, D_4 consist of the symmetries of the equilaterial triangle and the square, respectively. In a similar way, the Klein 4-group $\mathcal{V}$ can be viewed as group of symmetries of the non-square rectangle $\Box ABCD$ (Figure 1.5). Elements of $\mathcal{V}$ consist of the two clockwise rotations of $0°$ and $180°$ about the center O of rectangle $\Box ABCD$, denoted as $e = \rho_0$ and $a = \rho_2$, together with the two reflections through the perpendicular lines ℓ_3, ℓ_4, denoted by $b = \sigma_1$ and $c = \sigma_2$. With this interpretation, $\mathcal{V}$ is a subgroup of D_4.

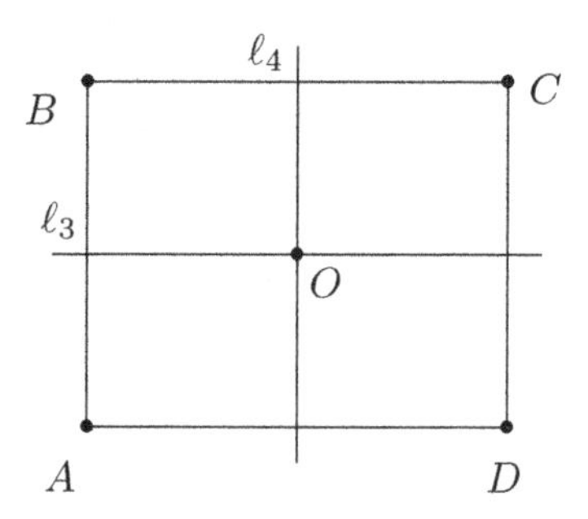

Fig. 1.5 Non-square rectangle $\Box ABCD$, $\ell_3 \perp \overleftrightarrow{DC}$, $\ell_4 \perp \overleftrightarrow{AD}$.

Much number theory results from the next proposition.

Proposition 1.31. *Every subgroup of a cyclic group is cyclic.*

Proof. Let $G = \langle a \rangle$ be cyclic, and let $H \leq G$. If $H = \{e\}$ then H is cyclic and the proposition is proved. So we assume that H has at least two elements e and $b \neq e$. Since b is non-trivial and $H \leq \langle a \rangle$, there exists a positive integer k so that $a^k \in H$. By the well-ordering principal for natural numbers, there exists a smallest positive integer m for which $a^m \in H$.

Let $h \in H$. Then $h = a^n$ for some integer n. Now by the division algorithm for integers, $n = mq + r$ for integers q, r with $0 \leq r < m$. Thus $h = (a^m)^q a^r$, and so, $a^r = h(a^m)^{-q} \in H$. But this says that $r = 0$ since m was chosen to be minimal. Consequently, $h = (a^m)^q \in \langle a^m \rangle$, and so $H = \langle a^m \rangle$. $\Box$

Let $\{n_1, n_2, \cdots, n_k\}$ be a finite set of integers, at least one of which is non-zero, and consider the subgroup $H = \langle n_1, n_2, \cdots, n_k \rangle$ of Z. By Proposition 1.31, H is cyclic, that is, H is of the form $\langle d \rangle$ for some integer d. We may assume $d > 0$. This integer d is **the greatest common divisor** of the set of integers $\{n_1, n_2, \ldots, n_k\}$, and is denoted by $\gcd(n_1, n_2, \cdots, n_k)$.

The greatest common divisor is an appropriate name: if $d = \gcd(n_1, n_2, \cdots, n_k)$, then $n_i \in \langle d \rangle, \forall i$, and so, $dk_i = n_i$ for some integer k_i. Consequently, d divides each n_i. Also, if c is a common divisor of the n_i, then $\langle d \rangle = \langle n_1, n_2, \ldots, n_k \rangle \subseteq \langle c \rangle$, thus c divides d; in this sense, d is the "greatest" common divisor. Conversely, if d is a common divisor of the n_i which is divisible by any other common divisor, then $d = \gcd(n_1, n_2, \ldots, n_k)$. The greatest common divisor is generalized to commutative rings with unity in §2.1.

Readers already know how to compute $\gcd(n_1, n_2)$ for integers n_1, n_2. For instance, to find $\gcd(1750, 1176)$ we perform a series of divisions:

$$1750 = 1176 \cdot 1 + 574,$$

$$1176 = 574 \cdot 2 + 28,$$

$$574 = 28 \cdot 20 + 14,$$

$$28 = 14 \cdot 2 + 0,$$

and the last non-zero residue 14 is the greatest common divisor.

Proposition 1.32 (Bezout's Lemma). *Let n_1, n_2 be integers with $d = \gcd(n_1, n_2)$. Then there exists integers x, y for which $n_1x + n_2y = d$.*

Proof. Since $\gcd(n_1, n_2) = d$, the cyclic subgroup $\langle n_1, n_2 \rangle \leq Z$ is generated by d. Thus there exists integers x, y so that $n_1x + n_2y = d$. $\square$

But how do we find x and y? Are they unique? The computation of x and y follows from the calculation of $\gcd(n_1, n_2)$. For instance, to find x and y for which $1750x + 1176y = 14$, we solve for 14, 28, and 574 in the equations above and substitute, thus:

$$\begin{aligned} 14 &= 574 - 28 \cdot 20 \\ &= 574 - (1176 - 574 \cdot 2) \cdot 20 \\ &= 1176 \cdot -20 + 574 \cdot 41 \\ &= 1176 \cdot -20 + (1750 - 1176 \cdot 1) \cdot 41 \\ &= 1750 \cdot 41 + 1176 \cdot -61, \end{aligned}$$

and so, $x = 41$, $y = -61$.

Integers m, n are **coprime** if $\gcd(m, n) = 1$. For $n \geq 1$, the number of integers m, $1 \leq m \leq n$, for which $\gcd(m, n) = 1$ is **Euler's function** $\varphi(n)$. For a prime number p, $\varphi(p) = p - 1$.

A common multiple of a set of integers $\{n_1, n_2, \cdots, n_k\}$, not all zero, is an integer $m \geq 0$ for which n_i divides m for $1 \leq i \leq k$. Let S be the set of all common multiples of $\{n_1, n_2, \ldots, n_k\}$. Then by the well-ordering principle S has a smallest element which is the **least common multiple** of $\{n_1, n_2, \ldots, n_k\}$, denoted by $[n_1, n_2, \ldots, n_k]$.

Proposition 1.33. *Let n_1, n_2 be integers, not both zero. Then*

$$[n_1, n_2] = \frac{|n_1 n_2|}{\gcd(n_1, n_2)}.$$

Proof. Let $d = \gcd(n_1, n_2)$. One has $\gcd(n_1/d, n_2/d) = 1$ and so,

$$[n_1/d, n_2/d] = \frac{|n_1 n_2|}{d^2},$$

by §1.6, Exercise 59 and §1.6, Exercise 60. Now,

$$d[n_1/d, n_2/d] = [n_1, n_2]$$

by §1.6, Exercise 61, and so,

$$[n_1, n_2] = \frac{|n_1 n_2|}{d}.$$

□

Proposition 1.34. *Let $G = \langle a \rangle$ be cyclic of order n and let $s \geq 1$ be an integer. Then the order of the cyclic subgroup of G generated by a^s is n/d where $d = \gcd(n, s)$.*

Proof. Note that n is the smallest positive integer for which $a^n = e$. The order of $\langle a^s \rangle$ is the smallest positive integer m for which $(a^s)^m = a^{sm} = e$. Note that $sm \geq n$, but even more: n must divide sm. For if not, there exist positive integers v, w with $sm = nv + w$ with $0 < w < n$. Thus $e = a^{sm} = a^{nv}a^w = a^w$, which contradicts the minimality of n.

Thus the order of $\langle a^s \rangle$ is the smallest positive integer m for which n divides sm. We compute the value of m as follows. Since n divides sm, sm is a common multiple of s and n, hence $sm \geq [s, n]$. By Proposition 1.33, $m \geq [s, n]/s = n/d$, where $d = \gcd(n, s)$. Now, $(n/d)s = ns/d = [n, s]$, which is a multiple of n, and so $(a^s)^{(n/d)} = e$. It follows that $m = n/d$. □

Corollary 1.2. *Let $G = \langle a \rangle$ be cyclic of order n, and let s be an integer. Then $G = \langle a^s \rangle$ if and only if $\gcd(n, s) = 1$.*

Proof. Exercise. □

The following is a classical theorem of number theory.

Proposition 1.35 (The Chinese Remainder Theorem). *Let* n_1, n_2 *be integers that satisfy* $n_1, n_2 > 0$ *and* $\gcd(n_1, n_2) = 1$ *and let* a_1, a_2 *be any integers. Then the system of congruences*

$$\begin{cases} x \equiv a_1 \, mod \, n_1 \\ x \equiv a_2 \, mod \, n_2 \end{cases}$$

has a unique solution modulo $n_1 n_2$.

Proof. Define a map $\phi : Z \to Z_{n_1} \times Z_{n_2}$ by the rule $\phi(a) = (a \bmod n_1, a \bmod n_2)$. Clearly, ϕ is a group homomorphism since

$$\begin{aligned} \phi(a+b) &= ((a+b) \bmod n_1, (a+b) \bmod n_2) \\ &= (a \bmod n_1 + b \bmod n_1, a \bmod n_2 + b \bmod n_2) \\ &= (a \bmod n_1, a \bmod n_2) + (b \bmod n_1, b \bmod n_2) \\ &= \phi(a) + \phi(b). \end{aligned}$$

Now, $\ker(\phi)$ is a cyclic subgroup of Z generated by a common multiple of n_1, n_2, which in fact is the least common multiple $[n_1, n_2]$. Since $\gcd(n_1, n_2) = 1$, $[n_1, n_2] = n_1 n_2$ by Proposition 1.33. By the First Isomorphism Theorem (Proposition 1.22) there is an injective group homomorphism $\psi : Z_{n_1 n_2} \to Z_{n_1} \times Z_{n_2}$ defined as $\psi(a \bmod n_1 n_2) = (a \bmod n_1, a \bmod n_2)$. Since $Z_{n_1 n_2}$ and $Z_{n_1} \times Z_{n_2}$ have the same number of elements, ψ is a isomorphism. The Chinese Remainder Theorem follows immediately. □

The given proof is non-constructive: it does not tell us how to find the unique solution to the system. To find the solution, use Bezout' Lemma to find x_1, x_2 so that $n_1 x_1 + n_2 x_2 = 1$. Then

$$(a_1 - a_2)(n_1 x_1 + n_2 x_2) = (a_1 - a_2),$$

and

$$x = a_1 + n_1 x_1 (a_2 - a_1) = a_2 + n_2 x_2 (a_1 - a_2),$$

is the unique solution modulo $n_1 n_2$.

Proposition 1.36. $Z_{mn} \cong Z_m \times Z_n$ *if and only if* $\gcd(m, n) = 1$.

Proof. Assume that $\gcd(m, n) = 1$. From the proof of the Chinese Remainder Theorem, we already have $Z_{mn} \cong Z_m \times Z_n$. For the converse, suppose there is an isomorphism $Z_{mn} \cong Z_m \times Z_n$. Then $[m, n] Z_{mn} = 0$, and so, $mn \leq [m, n]$. But since $[m, n] = mn/d$ where $d = \gcd(m, n)$, $d = 1$. □

The structure of the finite cyclic group Z_n can be determined by Proposition 1.36

Proposition 1.37. *Z_n decomposes as*

$$Z_n \cong Z_{p_1^{e_1}} \times Z_{p_2^{e_2}} \times \cdots \times Z_{p_k^{e_k}}$$

where $n = p_1^{e_1} p_2^{e_2} \cdots p_k^{e_k}$ is the factorization of n into powers of distinct primes.

Proof. Exercise. □

The structure of any finitely generated abelian group is determined as follows.

Proposition 1.38. *Let G be a finitely generated abelian group. Then*

$$G \cong Z_{p_1^{e_1}} \times Z_{p_2^{e_2}} \times \cdots \times Z_{p_k^{e_k}} \times \underbrace{Z \times Z \times \cdots \times Z}_{t},$$

where the primes p_i are not necessarily distinct, $e_i \geq 0$ are integers and $t \geq 0$ is an integer. If G is finite, $t = 0$ and $|G| = p_1^{e_1} p_2^{e_2} \cdots p_k^{e_k}$.

Proof. For a proof see [Lang (1984), Chapter I, §10]. □

1.5 The Sylow Theorems

Proposition 1.38 gives the structure finite abelian groups. We know much less about non-abelian groups. In this section we consider Cauchy's Theorem and Sylow's Theorems which tell us about the structure of arbitrary finite groups.

Let X be a set and let G be a group.

Definition 1.12. An **action of G on X** is a map $G \times X \to X$ which satisfies

(i) $ex = x$ for all $x \in X$,

(ii) $(gg')(x) = g(g'x)$ for all $x \in X$, $g, g' \in G$.

A set X endowed with an action of G is a **G-set**.

Example 1.1. Any group G is a G-set where the action $G \times G \to G$ is the binary operation on G.

Example 1.2. Let $H \leq G$. Then G is an H-set where the action $H \times G \to G$ is defined as $hg = hgh^{-1}$.

Example 1.3. Let $H \leq G$ and let L denote the collection of left cosets of H in G. Then L is a G-set with action $G \times L \to L$ defined as $g(aH) = (ga)H$ (one should verify that this action is well-defined on left cosets).

Example 1.4. Here is an example from linear algebra: Let V be a vector space over $\mathbb{R}$. Then V is an $\mathbb{R}^\times$-set, viewing $\mathbb{R}^\times$ as a multiplicative group, with action given as scalar multiplication.

Let X be a G-set and let $g \in G$. Let

$$X^g = \{x \in X : \; gx = x\}.$$

The subset X^g consists of the elements $x \in X$ fixed by g. For instance, for the G-set $X = G$ of Example 1.1, $G^e = G$ and $G^g = \emptyset$ if $g \neq e$. Let

$$X^G = \{x \in X : \; gx = x, \forall g \in G\},$$

X^G is the set of $x \in X$ that are fixed by all elements of G. In Example 1.2, for G abelian, $G^H = G$, as one can check.

Let X be a G-set and let $x \in X$. Let

$$G_x = \{g \in G : \; gx = x\}.$$

Proposition 1.39. *G_x is a subgroup of G.*

Proof. Let $a, b \in G_x$. Then

$$(ab)x = a(bx) = ax = x,$$

so that G_x is closed under the binary operation of G. Clearly, $e \in G_x$. Finally, for $a \in G_x$, $a^{-1}x = x$ since $ax = x$, thus $a^{-1} \in G_x$. □

For $x \in X$, the subgroup $G_x \leq G$ is the **isotropy subgroup** of x.

Let X be a G-set and let $x \in X$. The **orbit** of x in X under G is

$$Gx = \{gx : \; g \in G\}.$$

Proposition 1.40. *The collection of orbits Gx of x as x ranges over X forms a partition of X.*

Proof. (Sketch) Define a relation on X as follows: $x \sim y$ if and only if $y \in Gx$. Then as one can check, $\sim$ is an equivalence relation on X, where the equivalences classes are precisely the orbits in X. Consequently, the orbits form a partition of X, for details see [Rotman (2002), §1.3, Proposition 1.54]. □

Proposition 1.41. *Let G be a finite group. Let X be a G-set and let $x \in X$. Then $|Gx| = [G : G_x]$.*

Proof. Let $x_1 \in Gx$. Then $x_1 = g_1 x$ for some $g_1 \in G$. Let L be the collection of left cosets of G_x in G. Let $\phi : Gx \to L$ be defined by $\phi(x_1) = g_1 G_x$. Suppose that g is some other element of G with $x_1 = gx$. Then $(g^{-1}g_1)x = x$, hence $g_1^{-1}g \in G_x$, and so $gG_x = g_1 G_x$. Consequently, ϕ is well-defined, that is, ϕ is a function.

We show that ϕ is a 1-1 correspondence. Suppose that $\phi(x_1) = \phi(x_2)$. Then $g_1 G_x = g_2 G_x$ for some $g_2 \in G$ with $x_2 = g_2 x$, where $g_2 = g_1 g$ for some $g \in G_x$. Now,

$$x_2 = g_2 x = (g_1 g)x = g_1 x = x_1,$$

thus ϕ is 1-1. Next, let $g \in G_x \in L$. Put $y = gx$. Then $\phi(y) = gG_x$, so that ϕ is onto. $\square$

In the case that X is finite ($|X| < \infty$), there are a finite number of orbits $Gx_1, Gx_2, \ldots, Gx_s$ in X represented by a finite subset $\{x_1, x_2, \ldots, x_s\}$ of X. One has the formula

$$|X| = \sum_{i=1}^{s} |Gx_i|. \tag{1.3}$$

Observe that X^G is the collection of one-element orbits in X; let $r = |X^G|$. Then upon renumbering the x_i if necessary, formula (1.3) is

$$|X| = |X^G| + \sum_{i=r+1}^{s} |Gx_i|. \tag{1.4}$$

Proposition 1.42. *Let X be a finite G-set with G a finite group of order p^n. Then $|X| \equiv |X^G| \bmod p$.*

Proof. By Proposition 1.41, $|Gx_i|$ divides $|G| = p^n$ for $1 \le i \le s$, and so, p divides $|Gx_i|$ for $r+1 \le i \le s$. Thus p divides $|X| - |X^G|$. $\square$

We can now state and prove Cauchy's Theorem.

Proposition 1.43 (Cauchy's Theorem). *Let p be a prime number and let G be a finite group. If p divides $|G|$, then G contains a subgroup of order p.*

Proof. Let X be the set of all p-tuples $(g_1, g_2, \ldots, g_p)$, $g_i \in G$, for which $g_1 g_2 \cdots g_p = e$. Since $g_p = (g_1 g_2 \cdots g_{p-1})^{-1}$, an element of X is completely determined by choosing arbitrary elements $g_1, g_2, \ldots, g_{p-1}$ in G. Hence $|X| = |G|^{p-1}$, and so, p divides $|X|$.

Let $\sigma = (1, 2, \ldots, p)$ be the cycle in S_p of length p. Let $H = \langle \sigma \rangle$, so that $|H| = p$. Now X is an H-set with action defined as

$$\sigma^i(g_1, g_2, \ldots, g_p) = (g_{\sigma^i(1)}, g_{\sigma^i(2)}, \ldots, g_{\sigma^i(p)}),$$

for $0 \leq i \leq p-1$. By Proposition 1.42, $|X| \equiv |X^H| \bmod p$, hence p divides $|X^H|$, and so, X^H contains at least two elements. But X^H consists of precisely those p-tuples $(g_1, g_2, \ldots, g_p)$ in X for which $g_1 = g_2 = \cdots = g_p$. It follows that there exists a non-trivial element $g \in G$ for which $\underbrace{(g, g, \ldots, g)}_{p} \in X$. Thus $g^p = e$ and $\langle g \rangle$ is a subgroup of G of order p. □

Cauchy's Theorem will help us prove Sylow's First Theorem. But first, we need another application of Proposition 1.42.

Proposition 1.44. *Let G be a finite group and let H be a subgroup of G of order p^n. Let $N(H)$ be the normalizer of H (see §1.6, Exercise 27). Then $[N(H) : H] \equiv [G : H] \bmod p$.*

Proof. Let L be the set of left cosets of H in G, one has $|L| = [G : H]$. L is an H-set as in Example 1.3, and by Proposition 1.42,

$$[G : H] \equiv |L^H| \bmod p.$$

By definition,

$$L^H = \{gH : \ h(gH) = gH, \ \forall h \in H\},$$

that is, L^H consists of those left cosets $gH \in L$ for which $g^{-1}Hg = H$. Consequently, L^H consists of precisely the left cosets of H in $N(H)$. Thus $|L^H| = [N(H) : H]$. The result follows. □

Proposition 1.45 (First Sylow Theorem). *Let G be a finite group with $|G| = p^n m$, $n \geq 1$, $p \nmid m$. Then:*

(i) G contains a subgroup of order p^i, for $1 \leq i \leq n$.

(ii) If H is a subgroup of G of order p^i for $1 \leq i \leq n-1$, then H is a normal subgroup of a subgroup of order p^{i+1}.

Proof. (i) We use induction on i:

(The trivial case) By Cauchy's Theorem, G contains a subgroup of order p.

(The induction step) Suppose that G contains a subgroup H of order p^i for $1 \le i \le n-1$. By Proposition 1.44, $p \mid [N(H) : H]$, and since $H \triangleleft N(H)$, p divides $|N(H)/H|$. So by Cauchy's Theorem, $N(H)/H$ contains a subgroup K of order p. Let $\psi : N(H) \to N(H)/H$ be the canonical surjection. Then $\psi^{-1}(K)$ is a subgroup of G of order p^{i+1}.

For (ii), just note that the subgroup H of (i) is normal in $\psi^{-1}(K)$. □

A subgroup of order p^n in G given in Proposition 1.45(i) is called a **Sylow p-subgroup** of G.

Sylow stated two other theorems. We defer their proofs, but see [Rotman (2002), §5.2].

Proposition 1.46 (Second Sylow Theorem). *Let H_1, H_2 be two Sylow p-subgroups of the group G. Then H_1 and H_2 are congugate, that is, there exists $g \in G$ for which $gH_1g^{-1} = H_2$.*

Proposition 1.47 (Third Sylow Theorem). *Let G be a finite group and suppose that p divides $|G|$. Let s be the number of Sylow p-subgroups of G. Then $s \equiv 1 \bmod p$ and s divides $|G|$.*

1.6 Exercises

Exercises for §1.1

(1) Let S be a finite set with k elements. Compute the number of binary operations on the set S.
(2) How may semigroups with exactly two elements exist?
(3) Give an example of a semigroup that is not a monoid.
(4) Let S be a monoid with identity element e. Show that e is unique.
(5) Determine whether the following sets are groups under the indicated binary operations.

   (a) Z, together with ordinary subtraction
   (b) The subset of Z defined as $2Z = \{2m : m \in Z\}$, together with ordinary addition
   (c) $\mathbb{R}$, together with ordinary multiplication
   (d) $\mathbb{R}^+$, together with ordinary multiplication
   (e) $\mathbb{R}^\times$, together with ordinary multiplication

(f) $Z \times Z$, together with the binary operation on $Z \times Z$ defined by $((a, b), (c, d)) \mapsto (a, d)$, $\forall a, b, c, d \in Z$.

(6) Find an example of a monoid that is not a group.

(7) Let D_4 denote the 4th dihedral group.

(a) Compute the group product $\rho_1 \mu_1 \sigma_2$.
(b) Compute $(\rho_3 \sigma_1 \mu_2)^{-1}$.

(8) Let

$$\sigma = \begin{pmatrix} 1\,2\,3\,4\,5\,6 \\ 2\,5\,4\,6\,3\,1 \end{pmatrix}, \quad \tau = \begin{pmatrix} 1\,2\,3\,4\,5\,6 \\ 3\,2\,1\,4\,6\,5 \end{pmatrix}$$

be elements in S_6, the symmetric group on 6 letters.

(a) Compute σ^{-2}.
(b) Compute $\tau \circ \sigma^2$.
(c) Decompose σ into a product of transpositions. Is σ an even or odd permutation?

(9) Prove Proposition 1.5.

(10) **(Alternate Definition of a Group)**
Definition 1.4′ A **group** is a set G together with a binary operation $G \times G \to G$ for which

(i) the binary operation is associative,
(ii) there exists an element $e \in G$ for which $ea = a$, for all $a \in G$,
(iii) for each $a \in G$, there exists an element $c \in G$ for which $ca = e$.

Show that this definition of a group is equivalent to Definition 1.4.

Exercises for §1.2

(11) Find all of the subgroups of Z_{15}.
(12) Find all of the subgroups of $Z_3 \times Z_3$.
(13) Prove that $K \leq H$ and $H \leq G$ implies that $K \leq G$.
(14) Let $H \leq G$, $K \leq G$. Prove that $(H \cap K) \leq G$.
(15) Let G be a group and let H be a finite non-empty subset of G. Prove the following: H is a subgroup of G if and only if H is closed under the binary operation of G.
(16) Let H be a subgroup of a group G. Let g be an element of G that is not an element of H and let h be an element of H. Show that $gh \notin H$.

(17) Show that $H = \{0, 2, 4, 6\} \leq Z_8$. Compute all of the left cosets of H in Z_8.
(18) Verify that $H = \{\rho_0, \rho_1, \rho_2\} \leq D_3$. Show that the left cosets of H in D_3 form a partition of D_3.
(19) Prove Proposition 1.10.
(20) Let S_n denote the symmetric group on n letters. Let $H = \{\sigma \in S_n : \sigma(n) = n\}$. Prove that $H \leq S_n$. How many elements are in H?
(21) Let A_n denote the collection of all of the even permutations in S_n, $n \geq 2$. Prove that $A_n \lhd S_n$. (A_n is the **alternating subgroup** of S_n.)
(22) Assume that G is finite with $H \lhd G$. Show that $|G/H|$ divides $|G|$.
(23) Let $H \lhd G$ and suppose that aH and bH are left cosets of H in G. Define

$$aHbH = \{ahbh' : \ h, h' \in H\}.$$

Prove that $aHbH = abH$.
(24) Compute the group table for the quotient group Z_{15}/H where $H = \{0, 3, 6, 9, 12\}$.
(25) Compute the group table for the quotient group $Z/4Z$.
(26) Assume that $H \lhd G$.

(a) Prove that $eH = H$ is the identity element in the quotient group G/H.
(b) Prove that for all $aH \in G/H$, $(aH)^{-1} = a^{-1}H$.

(27) Let $H \leq G$. Let $N(H) = \{g \in G : \ gHg^{-1} = H\}$.

(a) Prove that $N(H)$ is a subgroup of G.
(b) Prove that $H \lhd N(H)$. In fact, show that $N(H)$ is the largest subgroup of G which H is normal in ($N(H)$ is the **normalizer** of H in G).

(28) Let $\mathbb{R}$ denote the group of real numbers under ordinary addition. Show that $Z \lhd \mathbb{R}$ and $\mathbb{Q} \lhd \mathbb{R}$. Describe the elements of the quotient group $\mathbb{R}/Z$. Is there any reasonable description of $\mathbb{R}/\mathbb{Q}$?
(29) Let $H \leq G$ with G abelian. Prove that G/H is abelian.
(30) Suppose $H \leq G$ with $[G : H] = 2$. Show that $H \lhd G$. Suppose that $[G : H] = 3$. Is it necessarily true that $H \lhd G$?
(31) Suppose that $K \lhd H$ and $H \lhd G$. Prove or disprove: $K \lhd G$.

Exercises for §1.3

(32) Prove Proposition 1.16.

(33) Show that the map $f : \mathbb{R} \to \mathbb{R}^+$ defined as $f(x) = e^x$ is a homomorphism of groups.
(34) Prove that the map $\phi : Z \to Z_n$ defined as $\phi(a) = a \bmod n$ is a group homomorphism.
(35) Prove that Z_4 is isomorphic to the subgroup $\{\rho_0, \rho_1, \rho_2, \rho_3\}$ of D_4.
(36) Find all of the homomorphisms $\phi : Z_p \to Z_p$.
(37) Find all of the homomorphisms $\phi : D_3 \to D_3$.
(38) Let $\phi : Z_6 \to Z_3$ be the map defined as $\phi(a) = a \bmod 3$ for all $a \in Z_6$.

  (a) Prove that ϕ is a group homomorphism.
  (b) Determine $\ker(\phi)$.

(39) Use the Universal Mapping Property for Kernels to prove that there exists a group homomorphism $Z/6Z \to Z_3$.
(40) Let $\phi : Z_6 \to Z_5$ be the map defined as $\phi(a) = a \bmod 5$. Determine whether ϕ is a group homomorphism.
(41) Determine the kernel of the homomorphism $\phi : Z \to Z/nZ$ given by $\phi(a) = a + nZ$.
(42) Determine the kernel of the homomorphism $\phi : GL_n(\mathbb{R}) \to \mathbb{R}^\times$ given by $\phi(A) = \det(A)$.
(43) Compute the kernel of the homomorphism $\phi : \mathbb{R}^+ \to \mathbb{R}$ defined by $\phi(x) = \ln(x)$.
(44) Suppose $\phi : G \to G'$ is an injective group homomorphism. Prove that $\ker(\phi) = \{e\}$.
(45) Let G be any group. Prove that $G/\{e\} \cong G$ and $G/G \cong \{e\}$.
(46) Suppose N, N' are two normal subgroups of G with $G/N \cong G/N'$. Does it follow that $N \cong N'$?
(47) Prove that $Z \ncong \mathbb{Q}$ as groups.
(48) Let H, K, J be subgroups of abelian group G and suppose that $K \leq H$, $J \cap H = \{0\}$ and $J \cap K = \{0\}$. Prove that $(HJ)/(KJ) \cong H/K$.
(49) Let $\phi : G \to G'$ be a surjection of abelian groups, and let H be a subgroup of G. Show that there is a surjection of groups $\overline{\phi} : G/H \to G'/\phi(H)$.
(50) **(Cayley's Theorem for Finite Groups)** Let G be a finite group of order n. Show that G is isomorphic to a subgroup of S_n.

Exercises for §1.4

(51) Let G be an infinite cyclic group. Prove that $G \cong Z$.
(52) Prove that the map $\phi : Z_n \to C_n$ defined as $i \mapsto g^i$ is an isomorphism of groups.

(53) Compute all of the group homomorphisms $\phi : Z \to Z$.
(54) Let p be prime. Find all of the generators for the cyclic group Z_{p^3}.
(55) Suppose Z_m has exactly $m-1$ generators. Prove that m is prime.
(56) Find a set of generators for the Klein 4-group $\mathcal{V}$.
(57) Show that $\mathcal{V}$ cannot be isomorphic to a subgroup of a cyclic group.
(58) Compute $d = \gcd(28, 124)$. Find integers x, y for which $28x + 124y = d$.
(59) Let $n_1, n_2, \ldots, n_k$ be a finite set of integers, not all zero, with least common multiple $[n_1, n_2, \ldots, n_k]$. Let m be a common multiple of $n_1, n_2, \ldots, n_k$. Prove that $[n_1, n_2, \ldots, n_k] \mid m$.
(60) Let $n_1, n_2, \ldots, n_k$ be a finite set of integers, not all zero, and let $d = \gcd(n_1, n_2, \ldots, n_k)$. Suppose that d' is a common divisor of $n_1, n_2, \ldots, n_k$. Prove that $d' \mid d$.
(61) Let $n_1, n_2, \ldots, n_k$ be a finite set of integers, not all zero, with least common multiple $[n_1, n_2, \ldots, n_k]$. Let $r \geq 0$ be an integer. Prove that $r[n_1, n_2, \ldots, n_k] = [rn_1, rn_2, \ldots, rn_k]$.
(62) Prove by example that $[n_1, n_2, n_3] \neq (n_1 n_2 n_3)/\gcd(n_1, n_2, n_3)$.
(63) Let

$$\sigma = \begin{pmatrix} 1 & 2 & 3 & 4 \\ 4 & 1 & 2 & 3 \end{pmatrix}, \quad \tau = \begin{pmatrix} 1 & 2 & 3 & 4 \\ 3 & 2 & 1 & 4 \end{pmatrix}$$

be elements in S_4, the symmetric group on 4 letters.

(a) Compute the order of the cyclic subgroup $\langle \sigma \rangle$.
(b) Compute the order of the subgroup $\langle \sigma, \tau \rangle$.

(64) Prove Proposition 1.28.
(65) Let G be any group and let C be the subset of G defined as

$$C = \{aba^{-1}b^{-1} : \ a, b \in G\}.$$

(a) Prove that $\langle C \rangle$ is a normal subgroup of G. ($\langle C \rangle$ is the **commutator subgroup** of G.)
(b) Show that $G/\langle C \rangle$ is abelian.

(66) Prove Corollary 1.2.
(67) Let C_{14} denote the cyclic group of order 14 generated by g. List all of the generators of C_{14}.
(68) Compute the number of elements in the subgroup $\langle (1, 2) \rangle$ of $Z_4 \times Z_6$.
(69) List the elements in the quotient group $(Z_2 \times Z_8)/\langle (0, 2) \rangle$.
(70) Compute the number of elements in the subgroup $\langle (1, 2), (2, 3) \rangle$ of $Z_4 \times Z_6$.

(71) Find the unique solution in Z_{4300} of the system of congruences

$$\begin{cases} x \equiv 30 \bmod 43 \\ x \equiv 87 \bmod 100. \end{cases}$$

(72) Find three abelian groups of order 8, no two of which are isomorphic. Find four groups of order 8, no two of which are isomorphic.

(73) Decompose the group Z_{40} into a product of cyclic groups.

(74) Suppose that there are x isomorphism classes of abelian groups of order n, n odd. Compute the number of isomorphism classes of abelian groups of order $2n$.

Exercises for §1.5

(75) Let X be a G-set. The group G is **transitive** on X if for each $x_1, x_2 \in X$, there exists an element $g \in G$ with $gx_1 = x_2$. Determine which G-sets in Example 1.1-Example 1.4 have groups that are transitive on X.

(76) Let X be a G-set. Let $N = \{g \in G : \ gx = x, \forall x \in X\}$.

  (a) Show that $N \lhd G$.
  (b) The group G acts **faithfully** on X if $N = \{e\}$. Determine which actions in Example 1.1-Example 1.4 are faithful.

(77) Compute all of the Sylow 2, 3-subgroups of D_3.

(78) Let G be a finite abelian group. Prove that G has exactly one Sylow p-subgroup for each prime p dividing $|G|$.

Chapter 2

Rings

In this chapter we introduce the concept of a ring and give some elementary examples such as Z, $\mathbb{Q}$, $\mathbb{R}$, $\mathbb{C}$ and Z_n, the ring of integers modulo n. The major theme of this chapter will be to discuss various ways to construct new rings from a given ring R, e.g., the ring of matrices $\mathrm{Mat}_n(R)$, the ring of polynomials in a set of variables $R[\{x_\alpha\}]$, the monoid ring RS, the group ring RG and the cartesian product of rings. We define rings with unity, unit elements of a ring and the notion of a field. We discuss divisibility, irreducibility, integral domains and unique factorization domains. We emphasize the polynomial ring $K[x]$ for K a field, and prove the Division Theorem, the Factor Theorem, Gauss' Lemma, and Eisenstein's Criterion.

We then consider the group of units of a ring $U(R)$, compute $U(Z_n)$, and prove Fermat's Little Theorem. We show that the group ring ZC_2 has only trivial units (as does ZC_3, $Z(C_2 \times C_2)$ and ZC_4), and show that any finite subgroup of $U(F)$, F a field, is cyclic. We then cover characters of a group and generalize to group representations.

Next we cover ideals in a ring, their arithmetic and the Chinese Remainder Theorem for rings. In a commutative ring R, we consider the ideal generated by a subset $S \subseteq R$, specializing to principal ideals. We discuss principal ideal domains (PIDs) and show that $F[x]$ is a PID whenever F is a field. We define maximal and prime ideals in a commutative ring with unity and use Zorn's Lemma to show that every commutative ring contains at least one maximal ideal.

We introduce Noetherian rings and show that these rings are equivalent to rings that satisfy the ascending chain condition (ACC) for ideals. From this we prove that every PID is a UFD. We then turn to quotient rings and ring homomorphisms. We give the analogs for rings of the First Isomorphism Theorem and the Universal Mapping Property for Kernels and develop the notion of the characteristic of a ring.

In the final two sections we return to the main theme of the chapter – to construct new rings from existing rings. One section concerns the localization $S^{-1}R$ of a ring R at a multiplicative subset S of R. This shows how a ring of fractions can be built from an existing ring. The final section considers absolute values on $\mathbb{Q}$ and discusses how the rationals $\mathbb{Q}$ can be completed with respect to the ordinary absolute value to yield the field of real numbers $\mathbb{R}$. We also complete $\mathbb{Q}$ with respect to the p-adic absolute value to yield the p-adic rationals $\mathbb{Q}_p$ and the p-adic integers $\mathbb{Z}_p$. These completions show how the field $\mathbb{Q}$ can be extended to include elements represented as infinite sums of rationals.

2.1 Introduction to Rings

In this section we introduce the concept of a ring, discuss some basic properties of rings and give some elementary examples such as Z, $\mathbb{Q}$ and Z_n, the ring of integers modulo n. Give a ring R, we show how to construct new rings including the ring of matrices $\mathrm{Mat}_n(R)$, the ring of polynomials $R[x]$, the monoid ring RS, the group ring RG and the cartesian product of rings. We define commutative rings with unity, unit elements of a ring and the notion of a field. We discuss divisibility, zero divisors and integral domains, as well as reducible and irreducible elements and unique factorization domains (UFDs).

* * *

In contrast to a group, a ring is a set together with two binary operations.

Definition 2.1. A **ring** is a set R together with two binary operations, addition, $+$ and multiplication, $\cdot$ that satisfy

(i) R is an abelian group under addition with identity element 0_R,

(ii) for all $a, b, c \in R$, $a \cdot (b \cdot c) = (a \cdot b) \cdot c$ (associative law),

(iii) for all $a, b, c \in R$, $a \cdot (b + c) = a \cdot b + a \cdot c$ and $(a + b) \cdot c = a \cdot c + b \cdot c$ (distributive law).

It is often more convenient to denote the multiplication $a \cdot b$ by ab.

Proposition 2.1. *Let R be a ring, and let $0 = 0_R$. Then for all $a, b \in R$,* *(i)* $a0 = 0 = 0a$;

(ii) $(-a)b = a(-b) = -(ab)$.

Proof. For (i) note that $a0 = a(0+0) = a0 + a0$, and so, $a0 = 0$. Similarly, $0a = (0+0)a = 0a + 0a$, so that $0 = 0a$. For (ii): $(-a)b + ab = (-a+a)b = 0b = 0$, and so, $(-a)b = -(ab)$. Likewise, $a(-b) + ab = a(-b+b) = a0 = 0$, thus, $a(-b) = -(ab)$. □

A ring R is **commutative** if $ab = ba$ for all $a, b \in R$.

The most familiar examples of rings are taken from the sets of numbers that you have seen in other mathematics courses: Z, $\mathbb{Q}$, $\mathbb{R}$, and $\mathbb{C}$ are rings under the usual addition and multiplication.

Here are some other examples.

Example 2.1. Let $n \geq 1$. The subset nZ of Z is a ring under the operations induced or "borrowed" from the ring Z.

Example 2.2. The collection of residue classes modulo n, Z_n is a ring under modular addition $+_n$ and modular multiplication $\cdot_n$ defined as

$$a \bmod n +_n b \bmod n = (a+b) \bmod n,$$

$$a \bmod n \cdot_n b \bmod n = (ab) \bmod n.$$

For example, Z_4 is a ring with binary operations given by the tables

$+_4$	0	1	2	3
0	0	1	2	3
1	1	2	3	0
2	2	3	0	1
3	3	0	1	2

$\cdot_4$	0	1	2	3
0	0	0	0	0
1	0	1	2	3
2	0	2	0	2
3	0	3	2	1

There are several ways to build new rings from existing rings.

Example 2.3. Let R be a ring and let $\text{Mat}_n(R)$ denote the collection of all $n \times n$ matrices with entries in R. Then $\text{Mat}_n(R)$ is a ring under matrix addition and multiplication.

Example 2.4. Let R be a ring and let x be an indeterminate. Then the collection of all polynomials

$$p(x) = a_0 + a_1x + a_2x^2 + \cdots + a_{n-1}x^{n-1} + a_nx^n, \quad a_i \in R,$$

is a ring under the usual polynomial addition and multiplication, thus:

$$\sum_{i=0}^{n} a_ix^i + \sum_{i=0}^{n} b_ix^i = \sum_{i=0}^{n}(a_i + b_i)x^i,$$

$$(\sum_{i=0}^{m} a_ix^i)(\sum_{j=0}^{m} b_jx^j) = \sum_{i=0}^{m}\sum_{j=0}^{n} a_ib_jx^{i+j}.$$

We denote this **ring of polynomials** by $R[x]$. If $a_n \neq 0$, then the **degree** of $p(x)$ is n. We assume that the degree of the zero polynomial $p(x) = 0$ is $-\infty$.

In Example 2.4, take R to be the polynomial ring $R[x]$ and let y be an indeterminate. Then we can form the ring of polynomials in y over $R[x]$, which is denoted as $R[x][y]$ (or alternatively, as $R[x, y]$). This is the ring of polynomials in two indeterminates x, y. In $R[x, y]$, x, y commute: $xy = yx$.

This construction can be extended: Let R be a ring and let $\{x_\alpha\}_{\alpha \in J}$ denote a set of indeterminates indexed by an arbitrary set J. Then the set of all polynomials in $\{x_\alpha\}_{\alpha \in J}$ over R is a ring with the obvious polynomial addition and multiplication–the x_α commute. This ring of polynomials is denoted as $R[\{x_\alpha\}]$. If $\{x_\alpha\}$ is finite we write $R[x_1, x_2, \ldots, x_n]$.

Example 2.5. Let R be a ring and let S be a monoid with identity element 1. Let RS denote the collection of all quantities of the form

$$\sum_{s \in S} a_ss, \quad a_s \in R,$$

where $a_s = 0$ for all but a finite number of indices s. Then RS is a ring with addition and multiplication defined as follows:

$$\sum_{s \in S} a_ss + \sum_{s \in S} b_ss = \sum_{s \in S}(a_s + b_s)s,$$

$$(\sum_{s \in S} a_ss)(\sum_{t \in S} b_tt) = \sum_{s \in S}\sum_{t \in S} a_sb_tst,$$

where st is the monoid product in S. This is the **monoid ring** RS **of** S **over** R.

In the monoid ring RS, the element $r1$, $r \in R$, is identified with the ring element r. If x is an indeterminate, then the set of powers $S = \{1, x, x^2, x^3, \ldots\}$ is a monoid. In this case the monoid ring RS is precisely the polynomial ring $R[x]$.

If R is a commutative ring, then by definition, elements of R commute with elements of S, and so $\alpha r = r\alpha$ for $r \in R$, $\alpha \in RS$.

A special case of the monoid ring is given in the next example.

Example 2.6. Let R be a ring and let G be a finite group of order n consisting of the elements $1 = g_0, g_1 \ldots, g_{n-1}$. Then the collection of all quantities of the form

$$\sum_{i=0}^{n-1} a_i g_i = a_0 g_0 + a_1 g_1 + a_2 g_2 + \cdots + a_{n-1} g_{n-1}, \quad a_i \in R,$$

is a ring under addition and multiplication defined as follows:

$$\sum_{i=0}^{n-1} a_i g_i + \sum_{i=0}^{n-1} b_i g_i = \sum_{i=0}^{n} (a_i + b_i) g_i,$$

$$(\sum_{i=0}^{n-1} a_i g_i)(\sum_{j=0}^{n-1} b_j g_j) = \sum_{i=0}^{n-1} \sum_{j=0}^{n-1} a_i b_j g_i g_j,$$

where $g_i g_j$ is the group product in G. This is the **group ring** RG **of** G **over** R.

For an example of a group ring, let $R = Z$ and let $C_2 = \{1, g\}$, $g^2 = 1$, denote the cyclic group of order 2. Then ZC_2 consists of all quantities of the form $a_0 + a_1 g$ where a_0, a_1 are integers. Thus in ZC_2, $(1+2g)(3-1g) = 1 + 5g$.

As with groups, we can form the direct product of rings.

Example 2.7. If R_i, for $i = 1, \ldots, k$, is a collection of rings, then the cartesian product $\prod_{i=1}^{k} R_i$ is a ring, where

$$(a_1, a_2, \ldots, a_k) + (b_1, b_2, \ldots, b_k) = (a_1 + b_1, a_2 + b_2, \ldots, a_k + b_k);$$

and

$$(a_1, a_2, \ldots, a_k)(b_1, b_2, \ldots, b_k) = (a_1 b_1, a_2 b_2, \ldots, a_k b_k).$$

To illustrate a direct product we consider the ring $Z \times Z$ where for instance, $(5,4)+(-3,1)=(2,5)$ and $(3,2)(-1,6)=(-3,12)$.

Definition 2.2. A **ring with unity** is a ring R for which there exists a multiplicative identity element, denoted by 1_R and distinct from 0_R, that satisfies $1_R a = a 1_R = a$ for all $a \in R$.

As one can easily check, $Z \times Z$ is a ring with unity, where the unity element is $1_{Z \times Z} = (1,1)$.

If R is a ring with unity and RS is a monoid ring, then the element $1_R s$, $s \in S$ is identified with s. Consequently, in a monoid ring RS over a ring with unity,

$$1_R = 1_R 1 = 1$$

where 1 is the identity in the monoid S.

Definition 2.3. Suppose R is a ring with unity. An element $u \in R$ for which there exists an element $a \in R$ that satisfies $au = ua = 1_R$ is a **unit** of R. The element a is a **multiplicative inverse** of u. If u is a unit of R, then there is exactly one multiplicative inverse of u, denoted by u^{-1}.

For example, in the ring $\mathbb{Q}$, $1_{\mathbb{Q}} = 1$, and every non-zero element is a unit. In the ring $Z \times Z$, the only units are $(1,1)$, $(1,-1)$, $(-1,1)$, and $(-1,-1)$.

Definition 2.4. A **field** is a commutative ring with unity in which every non-zero element is a unit.

For example, the ring $\mathbb{Q}$ is a field, as is $\mathbb{R}$. Let F be a field and suppose that E is a field for which $F \subseteq E$. Then E is an **extension field** of F, which we denote by E/F. For instance, $\mathbb{R}$ is a field extension of $\mathbb{Q}$.

Definition 2.5. Let R be a commutative ring with unity and let $a, b \in R$. Then a **divides** b (denoted as $a \mid b$) if there exists an element $c \in R$ for which $b = ac$. If $b = ac$ with c a unit of R, then a and b are **associates**.

Definition 2.6. Let R be a commutative ring with unity and let $a, b \in R$. An element $d \in R$ is a **greatest common divisor** of a, b if

(i) $d \mid a$ and $d \mid b$;

(ii) $c \mid d$ whenever c is a common divisor of a and b.

Suppose $R = Z$ and let $m, n \in Z$. Then the $\gcd(m,n)$ defined in §1.4 is a greatest common divisor of m, n. In the group ring ZC_3, with $\langle g \rangle = C_3$,

$1 + g$ is a greatest common divisor of $1 + g$ and $1 - 2g - g^2$. In the field $\mathbb{Q}$ any non-zero element is a greatest common divisor of $1/2$ and -3.

In a ring, as we have seen, one has $0 = a0$ for $a \in R$. But suppose $0 = ac$ for $a \neq 0$, $c \in R$. Does it follow that $c = 0$? Not always. For example, in the ring Z_6, $2 \cdot 3 = 0$ with $2, 3$ non-zero; so there can be non-zero elements whose product is 0.

Definition 2.7. Let R be a ring and suppose that a, b are non-zero elements such that $ab = 0$. Then a, b are **zero divisors** in the ring R.

For example, we have $\begin{pmatrix} 1 & 0 \\ 0 & 0 \end{pmatrix} \begin{pmatrix} 0 & 0 \\ 0 & 1 \end{pmatrix} = \begin{pmatrix} 0 & 0 \\ 0 & 0 \end{pmatrix}$, and thus $\mathrm{Mat}_2(Z)$ has zero divisors. Moreover, $(2, 0)$ and $(0, -3)$ are zero divisors in $Z \times Z$. On the other hand, Z has no zero divisors.

Definition 2.8. An **integral domain** is a commutative ring with unity with no zero divisors.

For example, Z is an integral domain; if R is an integral domain, then so is $R[x]$.

Proposition 2.2. *Every field is an integral domain.*

Proof. Exercise. □

Proposition 2.3. *Every finite integral domain is a field.*

Proof. Let R be an integral domain with a finite number of elements. Necessarily, R is commutative. So it remains to show that every non-zero element of R is a unit. The elements of R may be listed as

$$a_0, a_1, a_2, \ldots, a_k,$$

where $a_0 = 0_R$ and $a_1 = 1_R$, and where $a_i \neq a_j$ if $i \neq j$. Multiplying this list on the left by an element $a \in R$, $a \neq 0$, yields

$$0, a, aa_2, \ldots, aa_k.$$

Now suppose $aa_i = aa_j$ for some i, j, $1 \leq i, j \leq k$. Then $aa_i + (-(aa_j)) = 0$. By Proposition 2.1(ii), $aa_i + a(-a_j) = a(a_i + (-a_j)) = 0$. Since R has no zero divisors, $a_i + (-a_j) = 0$, hence $a_i = a_j$. Thus $i = j$ which says that each aa_i is distinct. It follows that $aa_i = 1_R$ for some i. This shows that the non-zero element a has a right inverse a_i. Since R is commutative, a_i is also a left inverse. □

Let R be an integral domain. A non-zero, non-unit element $b \in R$ is **irreducible** if whenever $b = ac$, either a or c is a unit of R. A non-zero, non-unit element of R that is not irreducible is **reducible**. For example, a prime number p is irreducible in Z, but is a unit of $\mathbb{Q}$. The polynomial $x^2 - 2$ is irreducible in $Z[x]$, but reduces in $\mathbb{R}[x]$. If b is an irreducible element of R then ub is irreducible for all units $u \in R$.

Definition 2.9. An integral domain R is a **unique factorization domain (UFD)** if every non-zero, non-unit $r \in R$ can be written as a product of irreducibles

$$r = q_1 q_2 \cdots q_k$$

which is unique in the sense that if $s_1 s_2 \cdots s_l$ is any other factorization of r into a product of irreducibles, then $k = l$ and the s_i can be renumbered so that $q_i = u_i s_i, \forall i$, for units $u_1, \ldots, u_k \in R$.

For example, Z is a UFD; $F[x]$ is a UFD whenever F is a field.

2.2 Polynomial Rings

The section concerns the ring of polynomials over a field F. We prove the Division Theorem, the Factor Theorem, and show that every polynomial of degree $n \geq 1$ has at most n zeros in F. We specialize to $\mathbb{Q}[x]$, discuss primitive polynomials and prove Gauss' Lemma, which leads to a proof of Eisenstein's Criterion. As an application, we show that $p(x) = x^{p-1} + x^{p-2} + \cdots + x^2 + x + 1$ is irreducible over $\mathbb{Q}$. Consequently, $x^p - 1 = (x-1)p(x)$ is the factorization of $x^p - 1$ into a product of monic irreducible polynomials, and so (as we will see in Chapter 4) the minimal polynomial of a primitive pth root of unity is $p(x)$.

* * *

Let F be a field and let $F[x]$ be the ring of polynomials over F.

Proposition 2.4 (Division Theorem). *Let $f(x)$ and $g(x)$ be polynomials in $F[x]$ with $f(x) \neq 0$. Then there exist unique polynomials $q(x)$ and $r(x)$ for which*

$$g(x) = f(x)q(x) + r(x)$$

where $deg(r(x)) < deg(f(x))$.

Proof. Our proof is by induction on the degree of $g(x)$. If $\deg(g(x)) < \deg(f(x))$, then $g(x) = f(x) \cdot 0 + g(x)$, as required by the Division Theorem. So we assume that $\deg(g(x)) \geq \deg(f(x))$. Let $f(x) = a_0 + a_1 x + \cdot + a_m x^m$, $a_m \neq 0$, $g(x) = b_0 + b_1 x + \cdots + b_{m+n} x^{m+n}$, $b_{m+n} \neq 0$, $n \geq 0$. For the trivial case, suppose that $\deg(g(x)) = 0$. Then $g(x) = c$, $f(x) = d$, with c, d non-zero, and so, $c = (c/d)d + 0$, which proves the theorem.

Next, put

$$h(x) = g(x) - \left(\frac{b_{m+n}}{a_m}\right) x^n f(x).$$

Then $\deg(h(x)) < \deg(g(x))$, thus by the induction hypothesis, there exists $q_1(x)$, $r(x)$ so that

$$h(x) = f(x) q_1(x) + r(x),$$

with $\deg(r(x)) < \deg(f(x))$. Now,

$$g(x) = f(x)\left(q_1(x) + \left(\frac{b_{m+n}}{a_m}\right) x^n\right) + r(x),$$

which proves the theorem.

We leave it to the reader to prove the uniqueness of $q(x)$ and $r(x)$. □

Let $f(x) \in F[x]$. A **zero** (or **root**) of $f(x)$ in F is an element $a \in F$ for which $f(a) = 0$.

Proposition 2.5 (Factor Theorem). *Let $f(x) \in F[x]$ be a polynomial and let $a \in F$ be a zero of $f(x)$. Then $x - a$ divides $f(x)$.*

Proof. By the Division Theorem, there exist polynomials $q(x), r(x) \in F[x]$ for which

$$f(x) = (x - a) q(x) + r(x),$$

with $\deg(r(x)) < \deg(x - a) = 1$, and so, $r(x) = r$ for some $r \in F$. Now,

$$0 = f(a) = (a - a) q(a) + r = 0 \cdot q(a) + r = r,$$

and so, $r(x) = 0$, and consequently, $x - a \mid f(x)$. □

Proposition 2.6. *Let $f(x) \in F(x)$ be a polynomial of degree $n \geq 0$. Then $f(x)$ can have at most n roots in F.*

Proof. If $f(x)$ has degree 0, then $f(x) = c$, $c \neq 0$, and so, $f(x)$ has no zeros in F. So we assume that $\deg(f(x)) = n \geq 1$. If $a_1 \in F$ is a zero of $f(x)$, then by the Factor Theorem,

$$f(x) = (x - a_1)g_1(x),$$

where $g_1(x)$ is a polynomial in $F[x]$ of degree $n - 1$. If $a_2 \in F$ is a zero of $g_1(x)$, then again by the Factor Theorem

$$f(x) = (x - a_1)(x - a_2)g_2(x),$$

with $g_2(x) \in F[x]$ with $\deg(g_2(x)) = n - 2$. We continue in this manner until we arrive at the factorization

$$f(x) = (x - a_1)(x - a_2)(x - a_3) \cdots (x - a_k)g_k(x),$$

where $g_k(x)$ has no roots in F and $\deg(g_k(x)) = n - k$. Now, $a_1, a_2, \ldots, a_k$ is a list of k zeros of $f(x)$ in F. Note that $k \leq n$ since the degrees on the left hand side and the right hand side must be equal.

We claim that the a_i are all of the zeros of $f(x)$ in F. By way of contradiction, suppose that $b \in F$ satisfies $b \neq a_i$ for $1 \leq i \leq k$, with $f(b) = 0$. Then

$$0 = f(b) = (b - a_1)(b - a_2)(b - a_3) \cdots (b - a_k)g_k(b),$$

with none of the factors on the right hand side equal to 0. Thus F has zero divisors, a contradiction. □

We specialize to the field $F = \mathbb{Q}$ and consider the integral domain $\mathbb{Q}[x]$. For a given polynomial $f(x) \in \mathbb{Q}[x]$, it is often a challenge to determine whether $f(x)$ is irreducible or not. In what follows, we give a test for the irreducibility of $f(x)$ first published by Eisenstein in 1850.

A polynomial $f(x) \in \mathbb{Q}[x]$ is **primitive** if $f(x) \in Z[x]$ and the greatest common divisor of the coefficients of $f(x)$ is 1. Given $f(x) \in \mathbb{Q}[x]$ there is an element $r \in \mathbb{Q}$ so that $rf(x)$ is a primitive polynomial in $\mathbb{Q}[x]$.

The next proposition says that if a polynomial over Z has a factorization over $\mathbb{Q}$, then it can be factored over Z.

Proposition 2.7 (Gauss' Lemma). *Let $f(x) \in Z[x]$. Suppose that $f(x) = p(x)q(x)$ where $p(x), q(x) \in \mathbb{Q}[x]$. Then there exists polynomials $\hat{p}(x), \hat{q}(x) \in Z[x]$ so that $f(x) = \hat{p}(x)\hat{q}(x)$. Moreover, there exists elements $s, t \in \mathbb{Q}$ with $\hat{p}(x) = sp(x)$ and $\hat{q}(x) = tq(x)$.*

Proof. First note that there is an element $r \in \mathbb{Q}$ for which $g(x) = rf(x)$ is primitive. Since $f(x) \in Z[x]$, $1/r$ is an integer. We have

$$g(x) = rf(x) = rp(x)q(x).$$

Now, there exist elements $u, v \in \mathbb{Q}$ so that $urp(x)$ and $vq(x)$ are primitive. Consequently,

$$urp(x)vq(x) = uvrp(x)q(x) = uvg(x)$$

is primitive. Since $g(x)$ is primitive, $uv = \pm 1$. Thus $g(x) = \pm urp(x)vq(x)$ and so, $f(x) = \pm up(x)vq(x)$ with $\pm up(x)$, $vq(x)$ both in $Z[x]$. Setting $\hat{p}(x) = \pm up(x)$, $\hat{q}(x) = vq(x)$, $s = \pm u$, $t = v$ proves the result. □

With the help of Gauss' Lemma, we can establish the following test for irreducibility.

Proposition 2.8 (Eisenstein's Criterion). *Let*

$$f(x) = a_n x^n + a_{n-1} x^{n-1} + \cdots + a_2 x^2 + a_1 x + a_0$$

be a polynomial of degree $n \geq 1$ with coefficients in Z. Suppose that there exists a prime p for which $a_n \not\equiv 0 \bmod p$, $a_i \equiv 0 \bmod p$, $0 \leq i \leq n-1$, and $a_0 \not\equiv 0 \bmod p^2$. Then $f(x)$ is irreducible over $\mathbb{Q}$.

Proof. Suppose that $f(x)$ is reducible, that is, suppose there exist polynomials $p(x)$, $q(x)$ over $\mathbb{Q}$ with $\deg(p(x)) = k$, $\deg(q(x)) = l$, $l \geq k \geq 1$, for which $f(x) = p(x)q(x)$. Then by Gauss' Lemma there exist polynomials $\hat{p}(x)$, $\hat{q}(x)$ over Z with $\deg(\hat{p}(x)) = k$, $\deg(\hat{q}(x)) = l$ for which $f(x) = \hat{p}(x)\hat{q}(x)$. Write

$$\hat{p}(x) = b_k x^k + b_{k-1} x^{k-1} + \cdots + b_2 x^2 + b_1 x + b_0,$$

$$\hat{q}(x) = c_l x^l + c_{l-1} x^{l-1} + \cdots + c_2 x^2 + c_1 x + c_0,$$

for $b_i, c_j \in Z$. Equating coefficients in the factorization $f(x) = \hat{p}(x)\hat{q}(x)$ yields

$$\begin{aligned}
a_0 &= b_0c_0 \\
a_1 &= b_1c_0 + b_0c_1 \\
a_2 &= b_2c_0 + b_1c_1 + b_0c_2 \\
&\vdots \\
a_k &= b_kc_0 + b_{k-1}c_1 + \cdots + b_0c_k \\
a_{k+1} &= b_kc_1 + b_{k-1}c_2 + \cdots + b_0c_{k+1} \\
&\vdots \\
a_l &= b_kc_{l-k} + b_{k-1}c_{l-k+1} + \cdots + b_0c_l \\
a_{l+1} &= b_kc_{l-k+1} + b_{k-1}c_{l-k+2} + \cdots + b_0c_l \\
&\vdots \\
a_n &= b_kc_l.
\end{aligned}$$

Since $a_0 \not\equiv 0 \bmod p^2$ and $a_0 \equiv 0 \bmod p$, either $b_0 \equiv 0 \bmod p$ and $c_0 \not\equiv 0 \bmod p$ or $b_0 \not\equiv 0 \bmod p$ and $c_0 \equiv 0 \bmod p$. Suppose that $b_0 \equiv 0 \bmod p$ and $c_0 \not\equiv 0 \bmod p$. Then from the equations above, one obtains $b_i \equiv 0 \bmod p$ for $0 \leq i \leq k$, and so, $a_n \equiv 0 \bmod p$, a contradiction. Likewise, if $b_0 \not\equiv 0 \bmod p$ and $c_0 \equiv 0 \bmod p$, then we also obtain a contradiction. Thus no such factorization of $f(x)$ can exist; $f(x)$ is irreducible over $\mathbb{Q}$. □

Using the prime $p = 2$, one can easily show that $f(x) = x^n - 2$ is irreducible over $\mathbb{Q}$ for $n \geq 1$. Here is another application of Eisenstein's Criterion due to Gauss.

Proposition 2.9 (Gauss). *Let p be a prime number and let*

$$p(x) = x^{p-1} + x^{p-2} + \cdots + x^2 + x + 1.$$

Then $p(x)$ is an irreducible polynomial over $\mathbb{Q}$.

Proof. Let $q(x) = p(x+1)$. Then

$$\begin{aligned}
q(x) &= \frac{(x+1)^p - 1}{(x+1) - 1} \\
&= \frac{x^p + \binom{p}{1}x^{p-1} + \binom{p}{2}x^{p-2} + \cdots + \binom{p}{p-1}x}{x} \\
&= x^{p-1} + \binom{p}{1}x^{p-2} + \binom{p}{2}x^{p-3} + \cdots + \binom{p}{p-1}.
\end{aligned}$$

Now, for $1 \leq i \leq p-1$, $\binom{p}{i} \equiv 0 \bmod p$ with $\binom{p}{p-1} = p$. Thus $q(x)$ is irreducible over $\mathbb{Q}$ using the Eisenstein Criterion with the prime p. Now suppose $p(x)$ had a factorization $p(x) = f(x)g(x)$ with both $f(x)$ and $g(x)$ non-units in $\mathbb{Q}[x]$. Then

$$q(x) = p(x+1) = f(x+1)g(x+1)$$

with $f(x+1)$, $g(x+1)$ non-units in $\mathbb{Q}[x]$, thereby contradicting the irreducibility of $q(x)$. □

2.3 The Group of Units of a Ring

In this section we introduce the group of units $U(R)$ in a ring with unity, R. We compute the group of units of Z_n and show that $U(Z_n)$ has order equal to Euler's function $\varphi(n)$. As a corollary, we prove Fermat's Little Theorem. We next focus on $U(ZG)$, for G a finite group. We show that ZC_2 has only trivial units, as does ZC_3 and ZC_4. (In §4.3 we will prove that ZC_3 has only trivial units and that ZC_p for primes $p \geq 5$ has non-trivial units.) We next establish that if F is a field, then a finite subgroup of $U(F) = F^\times$ is cyclic. This fact leads to a discussion of characters of a finite group, which we generalize to the notion of group representations.

* * *

Throughout this section, we assume that R is a ring with unity 1_R. This implies that R has at least one other element which is necessarily the additive identity 0_R. When there is no chance of confusion, we will write these elements using the simpler notation: 0, 1.

A ring is provided with two binary operations, addition and multiplication. Under the addition, the ring is an abelian group. Is there a group arising from the multiplication of R?

Proposition 2.10. *Let R be a ring with unity. Let U denote the subset of elements that are units of R. Then U together with the ring multiplication is a group.*

Proof. This is a matter of showing that U is closed under the ring multiplication. Let u, v be units of R. Then since $v^{-1}u^{-1}uv = uvv^{-1}u^{-1} = 1$, uv is a unit with inverse $v^{-1}u^{-1}$. □

The group U in Proposition 2.10 is the **group of units** of R, and is denoted by $U(R)$. We observe that if R is a commutative ring, then $U(R)$ is an abelian group. For example, the group of units $U(Z)$ is the abelian group $\{1, -1\}$.

The group of units of the cartesian product of rings can easily be found.

Proposition 2.11. *Let $\prod_{i=1}^{k} R_i$ be the direct product of rings with unity. Then $U(\prod_{i=1}^{k} R_i) = \prod_{i=1}^{k} U(R_i)$.*

Proof. Exercise. □

We next compute $U(Z_n)$.

Proposition 2.12. *The group of units $U(Z_n)$ consists of all residue classes $1 \leq m \leq n-1$ for which $\gcd(m,n) = 1$. Thus $|U(Z_n)| = \varphi(n)$, where φ is Euler's function.*

Proof. We show that m is a unit of Z_n if and only if $\gcd(m,n) = 1$. Suppose $m \in Z_n$ with $\gcd(m,n) = 1$. Then by the definition of greatest common divisor, $\langle m,n \rangle = \langle 1 \rangle = Z$. Therefore, there exist integers x, y so that $mx + ny = 1$. Thus $mx \equiv 1 \mod n$. Let $x' = x \bmod n$. Then $x'm = mx' = 1$ in Z_n, so that $x' = m^{-1}$. Thus m is a unit in Z_n; we have also shown that $\varphi(n) \leq |U(Z_n)|$.

Now suppose that $m \in U(Z_n)$. Then there exists an integer $a \in Z$ for which $ma \equiv 1 \bmod n$, that is, there exists an integer y so that $ma + ny = 1$. This says that the subgroup $\langle m,n \rangle \leq Z$ is Z, thus $\gcd(m,n) = 1$. Now

$$|U(Z_n)| \leq \varphi(n) \leq |U(Z_n)|$$

which yields $|U(Z_n)| = \varphi(n)$. □

For example, $U(Z_{10}) = \{1,3,7,9\}$. Observe that $\langle 3 \rangle = U(Z_{10})$, so that $U(Z_{10}) \cong Z_4$. (Here we have an isomorphism of a multiplicative group onto an additive group.)

The following is a classical result attributed to Fermat.

Proposition 2.13 (Fermat's Little Theorem). *Let p be a prime and let $a \in Z$ with $p \nmid a$. Then $a^{p-1} \equiv 1 \bmod p$.*

Proof. By Proposition 2.12, $|U(Z_p)| = \varphi(p) = p - 1$. Let $r = a \bmod p$. Since $p \nmid a$, $r \in U(Z_p)$. Thus by Proposition 1.27, $r^{p-1} \equiv 1 \bmod p$. Since $a^{p-1} \equiv r^{p-1} \bmod p$, the result follows. □

Corollary 2.1. *Let p be a prime and let $a \in Z$. Then $a^p \equiv a \bmod p$.*

Proof. Exercise. □

Let G be a finite group and let ZG be the group ring over Z. Then $\{\pm g\}$, $g \in G$, is a set of units in ZG, called the **trivial units of** ZG. When does ZG have only trivial units? In the case $G = C_2 = \{1, \sigma\}$, we have the following result.

Proposition 2.14. *The group ring* ZC_2 *has only trivial units* $\{\pm 1, \pm\sigma\}$.

Proof. Observe that $(\pm 1)^2 = 1$, and so, if $a + b\sigma \in U(ZC_2)$ then both $a + b$ and $a - b$ are units in Z. Thus $a - b = \pm 1$ and $a + b = \pm 1$, which implies that $a = \pm 1$ and $b = 0$ or $a = 0$ and $b = \pm 1$. Consequently, ZC_2 has only trivial units. □

G. Higman [Higman (1940), Theorem 6] has shown that if G is abelian then ZG has only trivial units if and only if G is a finite cartesian product of groups of order 2, 3, 4 or 6. Consequently, ZC_3, $Z(C_2 \times C_2)$, $Z(C_2 \times C_3)$, ZC_4, and $Z(C_4 \times C_4)$ have only trivial units. On the other hand, ZC_5 has non-trivial units. For a well-written account of Higman's result, see [Sehgal (2013), Theorem 2.4]. In §4.3 we will show that ZC_3 has only trivial units and that ZC_p has non-trivial units for primes $p \geq 5$.

Now suppose we are given a ring which is a field F. Then most of the elements are units: the group of units $U(F)$ is $F^\times = F \backslash \{0\}$. Surprisingly, the structure of all finite subgroups of $U(F)$ is known.

Proposition 2.15. *Let* F *be a field, and let* G *be a finite subgroup of* $F^\times$. *Then* G *is cyclic.*

Proof. As a finite group, G is finitely generated, and since F is a commutative ring, G is abelian. Thus by Proposition 1.38 G decomposes as

$$G \cong Z_{d_1} \times Z_{d_2} \times \cdots \times Z_{d_k},$$

where d_j are powers of primes, not necessarily distinct.

Let $m = [d_1, d_2, \ldots, d_k]$. Then $m \leq d_1 d_2 \ldots d_k = |G|$. For each $a \in G$, $a^m = 1$, since $d_i \mid m$, for all $i = 1, \ldots, k$. Thus the elements of G are zeros of the polynomial $x^m - 1 \in F[x]$. By Proposition 2.6, $x^m - 1$ can have at most m zeros in F, thus $|G| \leq m \leq d_1 d_2 \ldots d_k = |G|$, which yields $[d_1, d_2, \ldots, d_k] = d_1 d_2 \ldots d_k$. This says that $\gcd(d_i, d_j) = 1$ for i, j, $i \neq j$. Thus by Proposition 1.37, $G \cong Z_m$ and so G is cyclic. □

Let $\mathbb{C}$ be the field of complex numbers and let H be a finite subgroup of $U(\mathbb{C}) = \mathbb{C}^\times$ of order m. Then by Proposition 2.15, H is cyclic. If h is a generator for H, then $h^{km} = 1$ for $k = 0, 1, \ldots, m-1$, and thus each element of H is a root of the polynomial $z^m - 1 \in \mathbb{C}[z]$. One root of this polynomial is the complex number

$$\zeta_m = \cos(2\pi/m) + \mathrm{i}\sin(2\pi/m),$$

while the other $m-1$ roots are powers of ζ_m:

$$\zeta_m^k = \cos(2\pi k/m) + \mathrm{i}\sin(2\pi k/m),$$

$k = 2, \ldots, m$. Together they constitute the m **mth roots of unity.** The group H is the **group U_m of the m mth roots of unity**, and is cyclic with generator ζ_m. By Corollary 1.2, any other generator for U_m is of the form ζ_m^k with $\gcd(k, m) = 1$. A generator for U_m is called a **primitive mth root of unity.** For example, in U_4, $\zeta_4 = \mathrm{i}$ and $\zeta_4^3 = -\mathrm{i}$ are the primitive 4th roots of unity. The primitive 5th roots of unity are ζ_5, ζ_5^2, ζ_5^3 and ζ_5^4.

Let G be a finite group. A **character of G** is a multiplicative group homomorphism $\gamma : G \to \mathbb{C}^\times$. Characters are important because they can be generalized to the notion of a group representation.

Let $G = C_p = \langle g \rangle$ be the cyclic group of order p, p prime. We determine all of the characters of C_p. Let $\gamma : C_p \to \mathbb{C}^\times$ be a character. By Lagrange's Theorem, the subgroup $\ker(\gamma) \le C_p$ is either trivial or all of C_p. If $\ker(\gamma) = C_p$, then $\gamma(g^j) = 1$ for $j = 0, \ldots, p-1$. This is the trivial character of C_p.

If $\ker(\gamma) = 1$, then $C_p \cong \gamma(C_p) \le \mathbb{C}^\times$ by Proposition 1.18, and so $\gamma(C_p) = U_p$. Therefore $\gamma(g) = \zeta_p^i$ for some i, $1 \le i \le p-1$, which is a primitive pth root of unity. Thus a non-trivial character is determined by sending g to one of the $(p-1)$ primitive pth roots of unity. Consequently, there are p characters of C_p consisting of the homomorphisms $\gamma_i : C_p \to \mathbb{C}^\times$ defined as $\gamma_i(g^j) = \zeta_p^{ij}$, for $0 \le i, j \le p-1$.

For example, consider $G = C_3 = \langle g \rangle$. Then there are 3 characters of C_3 given in the following tables.

x	$\gamma_0(x)$
1	1
g	1
g^2	1

x	$\gamma_1(x)$
1	1
g	ζ_3
g^2	ζ_3^2

x	$\gamma_2(x)$
1	1
g	ζ_3^2
g^2	ζ_3

The collection of characters of C_p, denoted by $\hat{C}_p$, forms a **group of characters** under the binary operation defined as

$$(\gamma_i\gamma_j)(g^k) = \gamma_i(g^k)\gamma_j(g^k) = \zeta_p^{k(i+j)}$$

for $0 \leq i, j, k \leq p-1$. Since γ_1 is so that $\gamma_1^i = \gamma_i$ for $0 \leq i \leq p-1$, $\hat{C}_p$ is cyclic of order p, and is generated by $\gamma = \gamma_1$; we have $\gamma^i(g^j) = \zeta_p^{ij}$ and γ^0 is the trivial character which we denote by 1. Clearly, $\hat{C}_p \cong C_p$, the isomorphism being given as $\gamma \mapsto g$.

As another example, we compute the character group of the Klein 4-group $\mathcal{V} = \{e, a, b, c\}$. Here $a^2 = b^2 = c^2 = e$ and $ab = c$, $ac = b$, $bc = a$. Now a homomorphism $\gamma : \mathcal{V} \to \mathbb{C}^\times$ must satisfy the relations $(\gamma(a))^2 = 1$, $(\gamma(b))^2 = 1$ and $(\gamma(c))^2 = 1$ in $\mathbb{C}^\times$. Thus the only possibilities for $\gamma(a)$, $\gamma(b)$ and $\gamma(c)$ are ± 1. Moreover, $\gamma(a)\gamma(b) = \gamma(c)$, $\gamma(a)\gamma(c) = \gamma(b)$ and $\gamma(b)\gamma(c) = \gamma(a)$. Thus the characters are as follows.

x	$\nu_1(x)$
e	1
a	1
b	1
c	1

x	$\nu_2(x)$
e	1
a	1
b	-1
c	-1

x	$\nu_3(x)$
e	1
a	-1
b	1
c	-1

x	$\nu_4(x)$
e	1
a	-1
b	-1
c	1

The characters of $\mathcal{V}$ form the character group $\hat{\mathcal{V}}$ under the product $(\nu_i\nu_j)(x) = \nu_i(x)\nu_j(x)$. We have $\hat{\mathcal{V}} \cong \mathcal{V}$.

Let G be a finite group and let $\gamma : G \to \mathbb{C}^\times$ be a character. The group $\mathbb{C}^\times$ can be identified with the elements in $\mathrm{GL}_1(\mathbb{C})$, the invertible 1×1 matrices with entries in $\mathbb{C}$. Thus the character γ can be written as the group homomorphism

$$\rho : G \to \mathrm{GL}_1(\mathbb{C}).$$

This is a 1-dimensional linear representation of G. Generalizing a bit, let G be any group and let K be a field. An **n-dimensional linear representation** of G is a group homomorphism

$$\rho : G \to \mathrm{GL}_n(K).$$

Linear representations of groups can be used to represent group elements as matrices so that the group operation can be represented by matrix multiplication. Representations of groups are important because they allow many group-theoretic problems to be reduced to problems in linear algebra which is well-understood. We give two examples of group representations (readers may need to recall some elementary linear algebra).

Example 2.8. Let KC_3 denote the group ring with $\langle g \rangle = C_3$. Then KC_3 is a vector space of dimension 3 over K on the basis $\{1, g, g^2\}$. It determines a 3-dimensional linear representation of C_3:

$$\rho : C_3 \to \mathrm{GL}_3(K)$$

as follows. For $i, j = 0, 1, 2$ define $\rho(g^i)$ to be the invertible 3×3 matrix over K whose jth column is the coordinate vector of $g^i g^j = g^{i+j}$ with respect to the basis $\{1, g, g^2\}$. Thus

$$\rho(1) = \begin{pmatrix} 1 & 0 & 0 \\ 0 & 1 & 0 \\ 0 & 0 & 1 \end{pmatrix}, \quad \rho(g) = \begin{pmatrix} 0 & 0 & 1 \\ 1 & 0 & 0 \\ 0 & 1 & 0 \end{pmatrix}, \quad \rho(g^2) = \begin{pmatrix} 0 & 1 & 0 \\ 0 & 0 & 1 \\ 1 & 0 & 0 \end{pmatrix}.$$

This is the **left regular** representation of C_3.

Example 2.9. Let D_3 denote the 3rd order dihedral group, the group of symmetries of the equilateral triangle consisting of 3 rotations ρ_0, ρ_1, ρ_2 and 3 reflections μ_1, μ_2, μ_3 (§1.1). The triangle is embedded in the xy-coordinate plane with origin $O = (0, 0)$ and vertices $A = (-\frac{\sqrt{3}}{2}, -\frac{1}{2})$, $B = (0, 1)$ and $C = (\frac{\sqrt{3}}{2}, -\frac{1}{2})$; the line ℓ_1 has equation $y = \frac{\sqrt{3}}{3}x$; ℓ_2 has equation $y = -\frac{\sqrt{3}}{3}x$, see Figure 2.1.

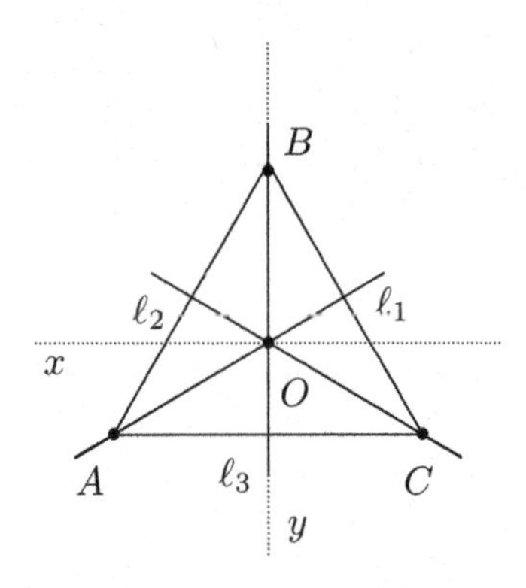

Fig. 2.1 Equilateral $\triangle ABC$ in the xy-plane.

We can represent D_3 as a group of 6 plane isometries, cf. [Martin (1982), Chapter 9]:

ρ_0 = rotation of a point (x, y) in the plane $0°$ clockwise about the origin O is given by the matrix multiplication

$$\begin{pmatrix} 1 & 0 \\ 0 & 1 \end{pmatrix} \begin{pmatrix} x \\ y \end{pmatrix} = \begin{pmatrix} x' \\ y' \end{pmatrix},$$

ρ_1 = rotation of a point (x, y) in the plane 120° clockwise about the origin O is given by the matrix multiplication

$$\begin{pmatrix} -1/2 & \sqrt{3}/2 \\ -\sqrt{3}/2 & -1/2 \end{pmatrix} \begin{pmatrix} x \\ y \end{pmatrix} = \begin{pmatrix} x' \\ y' \end{pmatrix},$$

ρ_2 = rotation of a point (x, y) in the plane 240° clockwise about the origin O is given by the matrix multiplication

$$\begin{pmatrix} -1/2 & -\sqrt{3}/2 \\ \sqrt{3}/2 & -1/2 \end{pmatrix} \begin{pmatrix} x \\ y \end{pmatrix} = \begin{pmatrix} x' \\ y' \end{pmatrix},$$

μ_1 = reflection of a point (x, y) in the plane through the line ℓ_1 is given by the matrix multiplication

$$\begin{pmatrix} 1/2 & \sqrt{3}/2 \\ \sqrt{3}/2 & -1/2 \end{pmatrix} \begin{pmatrix} x \\ y \end{pmatrix} = \begin{pmatrix} x' \\ y' \end{pmatrix},$$

μ_2 = reflection of a point (x, y) in the plane through the line ℓ_2 is given by the matrix multiplication

$$\begin{pmatrix} 1/2 & -\sqrt{3}/2 \\ -\sqrt{3}/2 & -1/2 \end{pmatrix} \begin{pmatrix} x \\ y \end{pmatrix} = \begin{pmatrix} x' \\ y' \end{pmatrix},$$

μ_3 = reflection of a point (x, y) in the plane through ℓ_3 (the y-axis) is given by the matrix multiplication

$$\begin{pmatrix} -1 & 0 \\ 0 & 1 \end{pmatrix} \begin{pmatrix} x \\ y \end{pmatrix} = \begin{pmatrix} x' \\ y' \end{pmatrix}.$$

The six matrices define a 2-dimensional representation of D_3

$$\rho : D_3 \to \mathrm{GL}_2(\mathbb{R}),$$

where, for example,

$$\rho(\rho_1) = \begin{pmatrix} -1/2 & \sqrt{3}/2 \\ -\sqrt{3}/2 & -1/2 \end{pmatrix}, \quad \rho(\mu_1) = \begin{pmatrix} 1/2 & \sqrt{3}/2 \\ \sqrt{3}/2 & -1/2 \end{pmatrix}.$$

It is routine to check that ρ is a group homomorphism.

2.4 Ideals

In this section we introduce ideals in a ring, show that the sum, product and intersection of ideals is an ideal, and apply ideals to the Chinese Remainder Theorem. Assuming that R is a commutative ring, we consider the ideal generated by a subset $S \subseteq R$, and specialize to finitely generated ideals, principal ideals, and principal ideal domains (PIDs). We prove that $F[x]$ is a PID whenever F is a field. We define maximal and prime ideals in a commutative ring with unity and use Zorn's Lemma to show that every commutative ring contains at least one maximal ideal. Finally, we introduce Noetherian rings and show that these rings are equivalent to rings that satisfy the ascending chain condition (ACC) for ideals. Using the ACC, we prove that every PID is a UFD.

* * *

By definition a ring is an abelian group under addition. The additive identity element is denoted by 0_R, or more simply 0. As an additive abelian group, R has at least two subgroups: $\{0\}$ and R. Some of the subgroups of R have a special property.

Definition 2.10. The additive subgroup $N \subseteq R$ is an **ideal** of R if $aN \subseteq N$ and $Nb \subseteq N$ for all $a, b \in R$.

If the ring R is commutative, then the definition of ideal is somewhat simpler: an additive subgroup $N \subseteq R$ is an ideal if $aN \subseteq N$ for all $a \in R$.

A ring R always has at least two ideals: the trivial ideal $\{0\}$ and the ring R itself. Here are some more examples. The subgroup nZ of Z is an ideal since $m(nk) = n(mk) \in nZ$ for all $m, n, k \in Z$. Let $R = Z[x]$ and let N be the subset of all polynomials which have constant term 0. Then N is an additive subgroup of R. Moreover, $p(x)q(x)$ has constant term 0 if $q(x)$ does, hence N is ideal of $Z[x]$. Here is an example of a subgroup which is not an ideal. Let $R = \mathbb{Q}$ with additive subgroup Z. Then Z fails to be an ideal of $\mathbb{Q}$ since $\frac{1}{2}Z \not\subseteq Z$.

Proposition 2.16. *Let R be a ring with unity. Let I be an ideal of R and let u be a unit in R. Then $uI = I$.*

Proof. By the definition of ideal, $uI \subseteq I$, so it remains to show that $I \subseteq uI$. Let $a \in I$. Then $a = (uu^{-1})a = u(u^{-1}a)$ with $u^{-1}a \in I$. Hence $a \in uI$. □

Proposition 2.17. *Let R be a ring with unity and suppose that I is an ideal of R which contains a unit. Then $I = R$.*

Proof. Since $I \subseteq R$ is immediate, we only need to show that $R \subseteq I$. Let u be a unit in I. Then $1 = u^{-1}u \in I$, and hence $r \cdot 1 \in I$ for all $r \in R$. $\square$

By Proposition 2.17, one sees that the only ideals of a field F are $\{0\}$ and F.

There is an arithmetic of ideals which we describe as follows. Let I and J be ideals of R. Define the **sum of ideals** $I + J$ to be the collection of all sums $a + b$, $a \in I$, $b \in J$, and define the **product of ideals** IJ to be the collection of all finite sums $\sum_{i=1}^{n} a_i b_i$ where $a_i \in I$, $b_i \in J$.

Proposition 2.18. *With the notation as above,*

(i) IJ *is an ideal,*

(ii) $I + J$ *is an ideal,*

(iii) $I \cap J$ *is an ideal.*

Proof. We prove (i), and leave (ii) and (iii) as exercises.

We first show that IJ is an additive subgroup if R. Note that $(\sum_{i=1}^{n} a_i b_i) + (\sum_{i=1}^{m} c_i d_i)$ is again a finite sum of the form $\sum ab$ with $a \in I$, $b \in J$. Also, 0 is a finite sum and $-\sum_{i=1}^{n} a_i b_i = \sum_{i=1}^{n}(-a_i)b_i$ is a finite sum. Thus $IJ \leq R$.

Now, $r(IJ) \subseteq IJ$, for all $r \in R$, since $r\sum_{i=1}^{n} a_i b_i = \sum_{i=1}^{n} r a_i b_i$ with $ra_i \in rI \subseteq I$, $b_i \in J$. Likewise, $(IJ)r \subseteq IJ$. Thus the product of ideals IJ is an ideal of R. $\square$

Let I be an ideal of R and let a, b be elements of R. We write $a \equiv b \bmod I$ if and only if $a - b \in I$. We have the following generalization of the Chinese Remainder Theorem.

Proposition 2.19 (Chinese Remainder Theorem for Rings). *Let R be a commutative ring with unity. Let N_1, N_2 be ideals of R with $N_1 + N_2 = R$ and let a_1, a_2 be elements of R. Then the system of congruences*

$$\begin{cases} x \equiv a_1 \bmod N_1 \\ x \equiv a_2 \bmod N_2 \end{cases}$$

has a unique solution modulo $N_1 N_2$.

Proof. There exist elements $n_1 \in N_1$, $n_2 \in N_2$ for which $n_1 + n_2 = 1$. Hence

$$(a_1 - a_2)(n_1 + n_2) = a_1 - a_2,$$

and so,

$$a_1 + n_1(a_2 - a_1) = a_2 + n_2(a_1 - a_2).$$

Thus $x = a_1 + n_1(a_2 - a_1)$ is a solution to the system of congruences. We show that this solution is unique modulo $N_1 N_2$. To this end, suppose that x' is another solution. Then $x - x' \equiv 0 \bmod N_1$ and $x - x' \equiv 0 \bmod N_2$, and so there exists an element $m \in N_1 \cap N_2$ for which $x - x' = m$. Now,

$$\begin{aligned} x - x' &= (x - x')(n_1 + n_2) \\ &= (x - x')n_1 + (x - x')n_2 \\ &= mn_1 + mn_2, \end{aligned}$$

with $mn_1 + mn_2 \in N_1 N_2$. It follows that $x' \equiv x \bmod (N_1 N_2)$. □

Corollary 2.2. *Let R be a commutative ring with unity and let I, J be ideals with $I + J = R$. Then $IJ = I \cap J$.*

Proof. Let $\sum_{i=1}^{n} a_i b_i \in IJ$. Then $a_i b_i \in I \cap J$ for each i, thus $IJ \subseteq I \cap J$. For the reverse containment, suppose that $a \in I \cap J$. Then $x = a$ is a solution to the system of congruences

$$\begin{cases} x \equiv 0 \bmod I \\ x \equiv 0 \bmod J \end{cases}$$

but so is $x' = 0$. Thus $a = a - 0 \in IJ$ by the Chinese Remainder Theorem for Rings. □

In a commutative ring, each subset of the ring gives rise to an ideal in a natural way.

Definition 2.11. Let $S = \{a_\beta\}$ be a subset of elements of the commutative ring R. Let N be the collection of all sums of the form $\sum_\beta r_\beta a_\beta$, where $r_\beta \in R$, and where $r_\beta = 0$ for all but a finite number of indices β. Then N is an ideal of R which we call the **ideal of R generated by S**.

If the generating set S is finite, then the ideal is **finitely generated**. If the ideal I is generated by $S = \{a_1, a_2, \ldots, a_k\}$, then I consists of all the linear combinations $\sum_{i=1}^{k} r_i a_i$, $r_i \in R$, and is denoted by $(a_1, a_2, \ldots, a_k)$.

Every ideal of a commutative ring is generated by some subset of the ideal. If necessary, one can take the ideal itself as the generating set. A challenging problem in ring theory is to find the smallest generating set for a given ideal. (This is analogous to extracting a basis from a spanning set for a vector space.) For example, the ideal $(2, x, x^2)$ of $Z[x]$ is the ideal $(2, x)$; the ideal $(4, 6)$ in Z is the ideal (2). Notice that $(4, 6) = (\gcd(4, 6))$.

When there is exactly one element in a generating set for an ideal, the ideal is given a special name.

Definition 2.12. Let R be a commutative ring and let $a \in R$. Then the ideal (a) is the **principal ideal generated by** a.

Not every ideal in a ring is principal, however. For example, the ideal $(2, x) \subseteq Z[x]$ cannot be written in the form $(p(x))$ for some $p(x) \in Z[x]$. To see this, suppose $(2, x) = (p(x))$ for some $p(x) \in Z[x]$. Then $2 = p(x)q(x)$ for some $q(x) \in Z[x]$, and hence $p(x)$ has degree 0, and is a constant which is necessarily ± 2. This says that $x \in (2) \subseteq Z[x]$, which is impossible.

Definition 2.13. A commutative ring R is a **principal ideal ring** if every ideal in R is principal.

Definition 2.14. An integral domain which is a principal ideal ring is a **principal ideal domain (PID)**.

Proposition 1.31 implies that Z is a PID by showing that every subgroup of a cyclic group is cyclic. By essentially the same argument as in Proposition 1.31, we can prove the following proposition.

Proposition 2.20. *Let F be a field. Then $F[x]$ is a PID.*

Proof. Clearly, $F[x]$ is an integral domain. Let I be a non-zero ideal of $F[x]$ (clearly, the zero ideal is principal). Let $p(x)$ be a non-zero polynomial of minimal degree in I. Then every element of I is a multiple of $p(x)$, for if $f(x)$ is in I, then by the Division Theorem for polynomials over a field, there exist polynomials $q(x)$ and $r(x)$ in $F[x]$ for which

$$f(x) = p(x)q(x) + r(x),$$

where $\deg(r(x)) < \deg(p(x))$. Thus $r(x) = f(x) - p(x)q(x) \in I$, and so $r(x) = 0$. $\square$

In a PID we have the following generalization of Bezout's Lemma.

Proposition 2.21. *Let R be a PID. Let $a, b \in R$, and suppose that d is* *a greatest common divisor of a, b. Then there exist elements $x, y \in R$ for which*

$$d = ax + by.$$

Proof. Let C be the set defined as

$$C = \{ar + bs : r, s \in R\}.$$

Then C is an ideal of R, and hence $C = (d')$ for some $d' \in R$. Now $d' \mid a$, $d' \mid b$ and $c \mid d'$ whenever $c \mid a$ and $c \mid b$, and so d' is a greatest common divisor of a, b.

Since d divides d', $d' = rd$ for some $r \in R$, and since d' divides d, $d = sd'$ for $s \in R$. Thus $d' = rsd'$. If $d' = 0$, then $d = 0$ and the proposition is proved. So we assume that $d' \neq 0$. Consequently, $rs = 1$, and so, r is a unit. By Proposition 2.16, $(d') = (d)$, and so $d = ax + by$ for some $x, y \in R$. $\square$

Definition 2.15. Let R be a commutative ring with unity. A **maximal ideal** is a proper ideal M of R for which there is no proper ideal N of R with $M \subset N \subset R$.

For example, the ideal (2) is a maximal ideal of Z, as is (3). However, since $(4) \subset (2) \subset Z$, (4) is not a maximal ideal. A commutative ring with unity with a unique maximal ideal is a **local ring**.

Definition 2.16. Let R be a commutative ring with unity. A proper ideal N is **prime** if $ab \in N$ implies that either $a \in N$ or $b \in N$.

For example, in Z, the ideal (5) is prime, but (6) is not prime.

How are maximal and prime ideals related? We know that there are non-maximal prime ideals, for example $\{0\}$ in Z. But we have the following proposition.

Proposition 2.22. *Every maximal ideal is prime.*

Proof. Suppose M is a maximal ideal with $ab \in M$. We show that either $a \in M$ or $b \in M$. By way of contradiction, we assume that $a, b \notin M$. Consider the ideals $(a) + M$ and $(b) + M$, which both contain M. Now if $(a) + M$ is not proper, then $(a) + M = R$, and

$$(b) + M = R((b) + M) = ((a) + M)((b) + M) \subseteq (ab) + M = M,$$

which says that $b \in M$.

If $(a)+M$ is proper, then $(a)+M = M$, thus $a \in M$. So M is prime. $\square$

It is somewhat surprising that every commutative ring with unity contains at least one prime ideal. This is proved using Zorn's Lemma which we briefly review.

Let S be a non-empty set. A **relation** on S is a subset $\preceq$ of the cartesian product $S \times S$. If $(x, y) \in \preceq$, we write $x \preceq y$. The relation $\preceq$ on S is **reflexive** if $x \preceq x, \forall x \in S$, $\preceq$ is **antisymmetric** if $x \preceq y$ and $y \preceq x$ implies $x = y$, and $\preceq$ is **transitive** if $x \preceq y$ and $y \preceq z$ implies that $x \preceq z$. We say that S is **partially ordered** under $\preceq$ if $\preceq$ is reflexive, antisymmetric and transitive. An element $m \in S$ is a **maximal element** if $x \in S$ and $m \preceq x$ implies that $m = x$. A subset T of a partially ordered set S is a **chain** if for all $x, y \in T$, either $x \preceq y$ or $y \preceq x$. An **upper bound** of a chain T is an element $u \in S$ for which $x \preceq u$ for all $x \in T$.

Zorn's Lemma. *Let S be a non-empty partially ordered set in which each chain has an upper bound. Then S has a maximal element.*

Proposition 2.23. *Let R be a commutative ring with unity. Then every proper ideal of R is contained in a prime ideal.*

Proof. Let J be a proper ideal of R and let $\mathcal{P}$ denote the collection of all proper ideals of R which contain J. Since $J \in \mathcal{P}$, $\mathcal{P}$ is a non-empty set which is partially ordered under set inclusion. Let $\mathcal{C}$ be any chain in $\mathcal{P}$. Then the ideal $\bigcup_{I \in \mathcal{C}} I$ is an upper bound for $\mathcal{C}$. Thus by Zorn's Lemma, $\mathcal{P}$ contains a maximal element M. By construction, M is a maximal ideal containing J. By Proposition 2.22, M is prime. $\square$

Since $\{0\}$ is a proper ideal of R, Proposition 2.23 shows that R has at least one prime ideal.

Corollary 2.3. *Let R be a commutative ring with unity. Suppose I is an ideal of R which is not contained in any prime ideal. Then $I = R$.*

Proof. This is just the contrapositive of Proposition 2.23. $\square$

In a UFD the prime ideals which are principal can be characterized.

Proposition 2.24. *Let R be a UFD, and let $a \in R$, $a \neq 0$. Then (a) is a prime ideal of R if and only if a is an irreducible element of R.*

Proof. Suppose (a) is prime with factorization $a = cb$. Then either $c \in (a)$ or $b \in (a)$. Suppose $c = ra$, $r \in R$. Then $a = rab = arb$, and so,

$a(1-rb) = 0$. Since $a \neq 0$ and R is an integral domain, $1 = rb$. Thus $b \in U(R)$ which says that a is irreducible.

Conversely, suppose that a is irreducible with $bc \in (a)$. Then $bc = ra$, for some $r \in R$. Let $b = b_1 b_2 \cdots b_k$, $c = c_1 c_2 \cdots c_l$, and $r = r_1 r_2 \cdots r_m$ be the unique factorizations of the elements b, c, r. Then,

$$(b_1 b_2 \cdots b_k)(c_1 c_2 \cdots c_l) = (r_1 r_2 \cdots r_m)a,$$

with $k+l = m+1$. Now the m factors of r on the right-hand side correspond to exactly $k+l-1$ factors on the left-hand side, which includes every factor of either b or c. Thus either $c \in (a)$ or $b \in (a)$, and so (a) is prime. □

Proposition 2.24 also holds in a PID.

Proposition 2.25. *Let R be a PID, and let $a \in R$, $a \neq 0$. Then (a) is a prime ideal of R if and only if a is an irreducible element of R.*

Proof. Exercise. □

Definition 2.17. Let R be a commutative ring with unity. Then R satisfies the **ascending chain condition for ideals (ACC)** if every ascending chain of ideals in R

$$I_0 \subseteq I_1 \subseteq I_2 \subseteq \cdots$$

eventually stops–that is, there exists an integer $m \geq 0$ for which $I_m = I_{m+1} = I_{m+2} = \cdots$.

The following was proved by E. Noether in 1921.

Proposition 2.26 (Noether). *Let R be a commutative ring with unity. The following are equivalent.*

(i) Every ideal I of R is finitely generated.

(ii) R satisfies the ACC.

Proof. (i) $\Longrightarrow$ (ii). Let

$$I_0 \subseteq I_1 \subseteq I_2 \subseteq \cdots$$

be an increasing sequence of ideals of R and let $I = \bigcup_{n=0}^{\infty} I_n$. Then I is an ideal of R, and so, I is finitely generated over R. Let $S = \{b_1, b_2, \ldots, b_l\}$ be a generating set for I. For $1 \leq i \leq l$, $b_i \in I_{m_i}$ for some integer m_i. Let $m = \max\{m_i\}$. Then $S \subseteq I_m$ and so,

$$I = I_m = I_{m+1} = I_{m+2} = \cdots .$$

(ii) $\Longrightarrow$ (i). Let I be an ideal of R and let $\mathcal{F}$ denote the collection of all finitely generated ideals of R that are contained in I. Certainly, $\mathcal{F}$ is non-empty since $\{0\} \in \mathcal{F}$. We claim that there exists a maximal element in $\mathcal{F}$, that is, we claim that there is an element $J' \in \mathcal{F}$ for which there is no $J \in \mathcal{F}$ with $J' \subset J$. To this end, assume that no such maximal element exists. Specifically, $I_0 = \{0\}$ is not maximal, thus there is an element $I_1 \in \mathcal{F}$ with $I_0 \subset I_1$. But I_1 is not maximal, and so there exists an element I_2 with $I_0 \subset I_1 \subset I_2$. Continuing in this manner, we construct an ascending chain of ideals that does not stop, violating the ACC. Consequently, the family $\mathcal{F}$ has a maximal element J. Note that J is finitely generated over R.

By construction, $J \subseteq I$. If $J \subset I$, then there is an element $x \in I \backslash J$. But then $J + (x)$ is an element of $\mathcal{F}$ with $J \subset J + (x)$, contradicting the maximality of J. Consequently, $J = I$ and we conclude that I is finitely generated. □

In honor of Noether, a commutative ring with unity that satisfies either of the two equivalent conditions of Proposition 2.26 is a **Noetherian ring**. For instance, any PID is Noetherian.

Proposition 2.27. *Let R be a Noetherian ring and let $\prod R$ denote the product of a finite number of copies of R. Then $\prod R$ is Noetherian.*

Proof. Exercise. □

Proposition 2.28 (Hilbert Basis Theorem). *Let R be a Noetherian ring. Then the polynomial ring $R[x]$ is Noetherian.*

Proof. We show that each ideal N of $R[x]$ is finitely generated as an R-module. For $n \geq 0$, let

$$J_n = \{a_n : \ a_n x^n + a_{n-1} x^{n-1} + \cdots + a_2 x^2 + a_1 x + a_0 \in N\}.$$

Then

$$J_0 \subseteq J_1 \subseteq J_2 \subseteq \cdots$$

is an increasing sequence of ideals of R. By Proposition 2.26, (i) $\Longrightarrow$ (ii), there exists an integer $m \geq 0$ for which

$$J_m = J_{m+1} = J_{m+2} = \cdots$$

For $n \geq 0$, let $\{b_{n,1}, b_{n,2}, \ldots, b_{n,j_n}\}$ be a generating set for J_n and let

$$\{f_{n,1}, f_{n,2}, \ldots, f_{n,j_n}\}$$

be a set of polynomials of degree n in N so that the leading coefficient of f_{n,i_n} is b_{n,i_n} for $1 \leq i_n \leq j_n$. We claim that the set

$$B = \bigcup_{n=0}^{m} \{f_{n,1}, f_{n,2}, \ldots, f_{n,j_n}\}$$

is a generating set for N as an $R[x]$-module. To prove the claim we proceed by induction on the degree of the polynomial in N. Clearly, a polynomial of degree ≤ 0 in N can be written as an $R[x]$-linear combination of elements in B. For the induction hypothesis, we assume that all polynomials of degree $\leq n-1$ can be written as $R[x]$-linear combinations of elements in B.

Let

$$f(x) = a_n x^n + a_{n-1}x^{n-1} + \cdots + a_2 x^2 + a_1 x + a_0$$

be a polynomial of degree n in N. There exists elements $r_1(x)$, $r_2(x), \ldots,$ $r_k(x) \in R[x]$ and elements $f_1(x)$, $f_2(x), \ldots, f_k(x) \in B$ so that

$$\begin{aligned} r_1(x)f_1(x) + r_2(x)f_2(x) + \cdots + r_k(x)f_k(x) \\ = a_n x^n + \text{terms of degree} \leq n-1. \end{aligned}$$

Let $g(x) = \sum_{t=1}^{k} r_t(x)f_t(x)$. Then $h(x) = f(x) - g(x)$ is an element of N of degree $\leq n-1$, which by the induction hypothesis is an $R[x]$-linear combination of elements in B. It follows that $f(x)$ is such a linear combination. □

Corollary 2.4. *Let K be a field and let $x_1, x_2, \ldots, x_n$ be indeterminates. Then $K[x_1, x_2, \ldots, x_n]$ is Noetherian.*

Proof. Since any field is Noetherian, $K[x_1]$ is Noetherian by Proposition 2.28. Another application of Proposition 2.28 yields $K[x_1][x_2] = K[x_1, x_2]$ Noetherian, and so on. □

Since every PID is Noetherian, every PID satisfies the ACC. This is the key to proving that every PID is a UFD.

Proposition 2.29. *Let R be a PID, and let a be a non-zero, non-unit of R. Then a is a product of irreducible elements of R.*

Proof. Suppose there exists a non-zero, non-unit element $a \in R$ which is not a product of irreducibles. Then a itself is not irreducible, and thus, $a = b_1 c_1$ where both b_1 and c_1 are non-units. Now either b_1 or c_1 cannot be written as a product of irreducibles, say $d_1 = b_1$ is one which can't be. Then d_1 is reducible and $d_1 = b_2c_2$ where neither b_2 nor c_2 is a unit. Now either b_2 or c_2 cannot be written as a product of irreducibles, let us say that $d_2 = b_2$ is this element. Note that $(d_1) \subset (d_2)$.

Now the element d_2 is reducible: $d_2 = b_3c_3$ where neither b_3 nor c_3 is a unit, and either b_3 or c_3 (assume b_3) cannot be written as a product of irreducibles. Set $d_3 = b_3$. Now $(d_1) \subset (d_2) \subset (d_3)$.

Continuing in this manner, one can construct a strictly ascending chain of ideals $\{(d_i)\}_{i=1}^{\infty}$. This is impossible by Proposition 2.26. □

Proposition 2.30. *If R is a PID, then R is a UFD.*

Proof. By Proposition 2.29, a non-zero, non-unit element of a PID can be factored into irreducibles, so we only need to show that this factorization is unique. Suppose that

$$p_1p_2\cdots p_l = q_1q_2\cdots q_m,$$

for irreducible elements p_i, q_j. By Proposition 2.25, (p_1) is prime and so, $q_j \in (p_1)$ for some j, $1 \le j \le m$. Consequently, $q_j = u_1p_1$ for $u_1 \in U(R)$. Upon renumbering the factors q_j if necessary, one has

$$p_1p_2\cdots p_l = u_1p_1q_2\cdots q_m,$$

thus

$$p_2p_3\cdots p_l = u_1q_2q_3\cdots q_m.$$

By similar reasoning $u_2p_2 = q_k$ for some unit $u_2 \in U(R)$ and some integer k, $2 \le k \le m$, and upon renumbering one obtains

$$p_2p_3\cdots p_l = u_1u_2p_2q_3q_2\cdots q_m,$$

thus

$$p_3\cdots p_l = u_1u_2q_3\cdots q_m.$$

Continuing in this manner yields $l = m$. It follows that the factorizations are unique in the sense of Definition 2.9. □

2.5 Quotient Rings and Ring Homomorphisms

In this section we define the quotient ring R/N of a ring R by an ideal N and show that R/N is a field if and only if N is a maximal ideal and that R/N is an integral domain if and only if N is prime. Next, we introduce the basic maps between two rings – ring homomorphisms, and give some examples of ring homomorphisms. We relate quotient rings and ring homomorphisms in the First Isomorphism Theorem (for rings) and the Universal Mapping Property for Kernels. As an application we develop the notion of the characteristic of a ring.

* * *

Let R be a ring and let N be an ideal of R. Since N is a normal subgroup of the additive group R, the quotient group R/N is defined with group operation

$$(a + N) + (b + N) = (a + b) + N. \tag{2.1}$$

One can endow R/N with the structure of a ring by defining a multiplication on the left cosets. For this we define a relation

$$B : R/N \times R/N \to R/N$$

by the rule $B(a + N, b + N) = ab + N$ for left cosets $a + N, b + N \in R/N$.

Proposition 2.31. *Let N be an ideal of R. Then the relation B defined above is a binary operation on R/N.*

Proof. We check that B is well-defined on left cosets. Suppose that $x \in a + N$, $y \in b + N$. Now $x = a + n_1$, $y = b + n_2$ for some $n_1, n_2 \in N$. Thus

$$xy = (a + n_1)(b + n_2) = ab + an_2 + n_1b + n_1n_2 \in ab + N,$$

and so,

$$B(x + N, y + N) = xy + N = ab + N = B(a + N, b + N).$$

□

Proposition 2.32. *Let N be an ideal of R. Then R/N is a ring with coset addition defined by (2.1) and coset multiplication defined as in Proposition 2.31.*

Proof. It is straightforward to show that R/N satisfies Definition 2.1. □

The ring R/N is the **quotient ring of R by N**. For an example, let $R = Z$, $N = 4Z$. Then the quotient ring $Z/4Z$ consists of the cosets $\{4Z, 1+4Z, 2+4Z, 3+4Z\}$, together with coset addition and multiplication. The binary operation tables for $Z/4Z$ are as follows

$+$	$4Z$	$1+4Z$	$2+4Z$	$3+4Z$
$4Z$	$4Z$	$1+4Z$	$2+4Z$	$3+4Z$
$1+4Z$	$1+4Z$	$2+4Z$	$3+4Z$	$4Z$
$2+4Z$	$2+4Z$	$3+4Z$	$4Z$	$1+4Z$
$3+4Z$	$3+4Z$	$4Z$	$1+4Z$	$2+4Z$

$\cdot$	$4Z$	$1+4Z$	$2+4Z$	$3+4Z$
$4Z$	$4Z$	$4Z$	$4Z$	$4Z$
$1+4Z$	$4Z$	$1+4Z$	$2+4Z$	$3+4Z$
$2+4Z$	$4Z$	$2+4Z$	$4Z$	$2+4Z$
$3+4Z$	$4Z$	$3+4Z$	$2+4Z$	$1+4Z$

We point out that the tables for $Z/4Z$ look just like the binary operation tables for the ring Z_4 given in §2.1.

For another example, let $R = \mathbb{Q}[x]$ and let $N = (x^2 - 2)$, the principal ideal of $\mathbb{Q}[x]$ generated by $x^2 - 2$. The elements of the quotient ring $\mathbb{Q}[x]/(x^2 - 2)$ consists of left cosets computed as follows. Let $f(x) \in \mathbb{Q}[x]$. By the Division Theorem there exists polynomials $q(x)$ and $r(x)$ for which

$$f(x) = q(x)(x^2 - 2) + r(x),$$

with $\deg(r(x)) < \deg(x^2 - 2) = 2$. Thus $r(x) = a + bx$ for some $a, b \in \mathbb{Q}$. It follows that the elements of $\mathbb{Q}[x]/(x^2 - 2)$ are $\{a + bx + (x^2 - 2) : \ a, b \in \mathbb{Q}\}$. Note that addition in $\mathbb{Q}[x]/N$ is given as

$$(a + bx + N) + (c + dx + N) = a + c + (b + d)x + N,$$

while multiplication is given as

$$\begin{aligned}(a + bx + N)(c + dx + N) &= (a + bx)(c + dx) + N \\ &= ac + (ad + bc)x + bdx^2 + N \\ &= ac + (ad + bc)x + 2bd - 2bd + bdx^2 + N \\ &= ac + 2bd + (ab + bc)x + bd(x^2 - 2) + N \\ &= ac + 2bd + (ab + bc)x + N.\end{aligned}$$

Let R be a ring with unity, 1. Then the quotient ring R/N is a ring with unity $1 + N$.

Proposition 2.33. *Let R be a commutative ring with unity, let N be a proper ideal of R and let $a \in R$. Then $a + N$ is a unit of R/N if and only if $(a) + N = R$.*

Proof. Suppose $(a) + N = R$. Since $1 \in R$, there exist elements $r \in R$ and $n \in N$ so that $ra + n = 1$, hence $ra = 1 - n$. Now

$$\begin{aligned}(r + N)(a + N) &= ra + N \\ &= (1 - n) + N \\ &= (1 + N) + (-n + N) \\ &= (1 + N) + N \\ &= 1 + N,\end{aligned}$$

thus $r + N = (a + N)^{-1}$.

Conversely, suppose $a + N$ is a unit of R/N. Then $1 + N = ar + N$ for some $r \in R$, and so, $1 \in ar + N$. Thus $R \subseteq (a) + N$. Since $(a) + N \subseteq R$, one has $(a) + N = R$. □

Proposition 2.34. *Let R be a commutative ring with unity. Then M is a maximal ideal of R if and only if R/M is a field.*

Proof. Suppose that R/M is a field and let N be a proper ideal of R with $M \subseteq N \subset R$. If $M \neq N$, then there exists an element $a \in N \backslash M$, and hence $a + M$ is a non-zero element of the field R/M. Consequently, $a + M$ is a unit in R/M, and so, by Proposition 2.33, $R = (a) + M \subseteq N$. Thus $N = R$, which is a contradiction.

For the converse, we suppose that M is maximal. Since R is a commutative ring with unity, so is R/M. So it remains to show that every non-zero element of R/M is a unit. Let $a + M \in R/M$, $a \notin M$. Then $(a) + M$ is an ideal of R with $M \subseteq (a) + M \subseteq R$. But M is maximal, so either $(a) + M = M$, or $(a) + M = R$. In the former case, $a \in M$, which is a contradiction. Thus $(a) + M = R$ which says that $a + M$ is a unit in R/M. □

Proposition 2.35. *Let R be a commutative ring with unity. Then N is a prime ideal of R if and only if R/N is an integral domain.*

Proof. Suppose that R/N is an integral domain, and let $ab \in N$. Then $N = ab + N = (a + N)(b + N)$, and so, either $a + N = N$, or $b + N = N$. Thus either $a \in N$ or $b \in N$, which says that N is prime.

For the converse, we suppose that N is prime. Since R is a commutative ring with unity, so is R/N. So it remains to show that R/N has no zero divisors. To this end, let $(a + N)(b + N) = N$. Then $ab + N = N$ so that $ab \in N$. Since N is prime, either $a \in N$ or $b \in N$, thus we must have either $a + N = N$, or $b + N = N$. □

When we considered functions from one group to another (§1.3), we were particularly interested in functions, namely group homomorphisms, that relate the group operations on the domain and codomain. Now we introduce functions from one ring to another that relate the two operations on each of the rings.

Definition 2.18. Let R, R' be rings. A map $\phi : R \to R'$ is a **ring homomorphism** if for all $a, b \in R$

(i) $\phi(a + b) = \phi(a) + \phi(b)$,

(ii) $\phi(ab) = \phi(a)\phi(b)$.

Definition 2.19. Let R, R' be rings with unity, with unity elements 1_R, $1_{R'}$. Then a map $\phi : R \to R'$ is a **homomorphism (of rings with unity)** if ϕ is a ring homomorphism and $\phi(1_R) = 1_{R'}$.

For example, the map $\phi : Z \to Z/nZ$ defined as $\phi(a) = a + nZ$ is a homomorphism of rings with unity. Indeed, we have already seen that ϕ is a homomorphism of groups, so (i) holds. For (ii), let $a, b \in Z$. Then $\phi(ab) = ab+nZ = (a+nZ)(b+nZ) = \phi(a)\phi(b)$. Moreover, $\phi(1) = 1+nZ = 1_{Z/nZ}$. Here is another example of a ring homomorphism.

Proposition 2.36. *Let E/F be a field extension and let $\alpha \in E$. Then the map $\phi_\alpha : F[x] \to E$ defined by $\phi_\alpha(p(x)) = p(\alpha)$ is a ring homomorphism.*

Proof. Let $p(x) = \sum_{i=0}^{m} a_i x^i$, $q(x) = \sum_{j=0}^{n} b_j x^j$ be polynomials in $F[x]$. Then

$$\begin{aligned}\phi_\alpha(p(x)+q(x)) &= \phi_\alpha(\sum_{i=0}^{m} a_i x^i + \sum_{j=0}^{n} b_j x^j) \\ &= \sum_{i=0}^{m} a_i \alpha^i + \sum_{j=0}^{n} b_j \alpha^j \\ &= \phi_\alpha(\sum_{i=0}^{m} a_i x^i) + \phi_\alpha(\sum_{j=0}^{n} b_j x^j)\end{aligned}$$

and

$$\begin{aligned}\phi_\alpha(p(x)q(x)) &= \phi_\alpha(\sum_{i=0}^{m}\sum_{j=0}^{n} a_i b_j x^{i+j}) \\ &= \sum_{i=0}^{m}\sum_{j=0}^{n} a_i b_j \alpha^{i+j} \\ &= \phi_\alpha(\sum_{i=0}^{m} a_i x^i)\phi_\alpha(\sum_{j=0}^{n} b_j x^j).\end{aligned}$$

Moreover, $\phi_\alpha(1_{F[x]}) = 1_F = 1_E$, and so ϕ_α is a homomorphism of rings with unity. □

The homomorphism of Proposition 2.36 is called the **evaluation homomorphism**. For an example, let $F = \mathbb{Q}$, $E = \mathbb{R}$, and $\alpha = \sqrt{2}$. Then the evaluation homomorphism $\phi_{\sqrt{2}} : \mathbb{Q}[x] \to \mathbb{R}$ is given by $p(x) \mapsto p(\sqrt{2})$.

Definition 2.20. The **kernel** of the ring homomorphism $\phi : R \to R'$ is the subset of R defined as $\ker(\phi) = \{a \in R : \phi(a) = 0\}$.

Proposition 2.37. *The kernel of a ring homomorphism $\phi : R \to R'$ is an ideal of R.*

Proof. Let $N = \ker(\phi)$. Then N is an additive subgroup of R, so we need only show that $aN \subseteq N$ and $Na \subseteq N$ for all $a \in R$. But these conditions follow since $\phi(an) = \phi(a)\phi(n) = 0$ for $an \in aN$, and $\phi(na) = \phi(n)\phi(a) = 0$ for $na \in Na$. □

Definition 2.21. A ring homomorphism $\phi : R \to R'$ is an **isomorphism** of rings if ϕ is a bijection. The rings R and R' are **isomorphic** if there exists an isomorphism $\phi : R \to R'$. We then write $R \cong R'$.

For example, the ring homomorphism $\phi : Z \times Z \to Z \times Z$ defined by $(m, n) \mapsto (n, m)$ is a ring isomorphism. Also, $\phi : Z/nZ \to Z_n$ given by $a + nZ \mapsto a \bmod n$ is a ring isomorphism.

A map can be a group isomorphism without being a ring isomorphism. For example, the map $\phi : Z \to 2Z$ given by $\phi(n) = 2n$ is an isomorphism if we consider Z and $2Z$ as groups, but it is not a ring homomorphism since $\phi(ab) = 2ab \neq (2a)(2b) = \phi(a)\phi(b)$.

A map, of course, can be a bijection of sets without being a ring isomorphism. There are many bijective maps between $\mathbb{R}$ and $\mathbb{C}$, but these rings are not isomorphic. To see this, we assume that there is an isomorphism (of commutative rings with unity) $\phi : \mathbb{R} \to \mathbb{C}$. Note that $\phi(1) = \phi(1_{\mathbb{R}}) = 1_{\mathbb{C}} = 1$. Since ϕ is a group homomorphism, $\phi(-1) = -\phi(1) = -1$. Since ϕ is surjective, there exists an element $r \in \mathbb{R}$ with $\phi(r) = \mathrm{i}$, thus $-1 = (\phi(r))^2 = \phi(r^2)$, and so, $\phi(-1) = \phi(r^2)$. Now since ϕ is an injection, $r^2 = -1$ for some real number r, which is impossible. Thus $\mathbb{R} \ncong \mathbb{C}$.

There is an elegant relationship between ring homomorphisms and quotient rings expressed in the following propositions (which are the analogs for rings of Proposition 1.21, Proposition 1.22 and Proposition 1.23).

Proposition 2.38. *Let N be an ideal of R. Then the map $\gamma : R \to R/N$ given by $\gamma(a) = a + N$ is a surjective ring homomorphism with kernel N.*

Proof. By Proposition 1.21, γ is a surjective homomorphism of additive groups with $\ker(\gamma) = N$. Now, $\gamma(ab) = ab + N = (a+N)(b+N) = \gamma(a)\gamma(b)$, which shows that γ is also a ring homomorphism. □

Proposition 2.39. *Let $\phi : R \to R'$ be a ring homomorphism with $\ker(\phi) = N$. Then $\gamma : R/N \to \phi(R)$ defined by $\gamma(a + N) = \phi(a)$ is a ring isomorphism.*

Proof. One first checks that $\phi(R)$ is a ring. By Proposition 1.22, γ is an additive group isomorphism with $\ker(\gamma) = \{N\}$. Now,

$$\begin{aligned}\gamma((a+N)(b+N)) &= \gamma(ab+N) \\ &= \phi(ab) \\ &= \phi(a)\phi(b) \\ &= \gamma(a+N)\gamma(b+N),\end{aligned}$$

so γ is also a ring homomorphism. □

Proposition 2.40 (Universal Mapping Property for Kernels). *Let $\phi : R \to R'$ be a ring homomorphism with $N = ker(\phi)$. Suppose that K is an ideal of R contained in N. Then there exists a surjective homomorphism of rings $\psi : R/K \to \phi(R)$ defined by $\gamma(a + K) = \phi(a)$.*

Proof. By the UMPK (Proposition 1.23) there exists a surjective homomorphism of groups $\psi : R/K \to \phi(R)$ defined by $\gamma(a + K) = \phi(a)$. Now, $\gamma((a + K)(b + K)) = \gamma(ab + K) = \phi(ab) = \phi(a)\phi(b) = \gamma(a + K)\gamma(b + K)$, so that γ is a ring homomorphism. □

Let $\phi : R \to R'$ be a ring homomorphism with $N = \ker(\phi)$. The Universal Mapping Property for Kernels (UMPK) says that given an ideal K of R with $K \subseteq N$, there exists a ring homomorphism $\psi : R/K \to R'$ so that

$$\psi s = \phi$$

where $s : R \to R/K$ is the canonical surjection; we say that ϕ "factors through" R/K and the following diagram commutes:

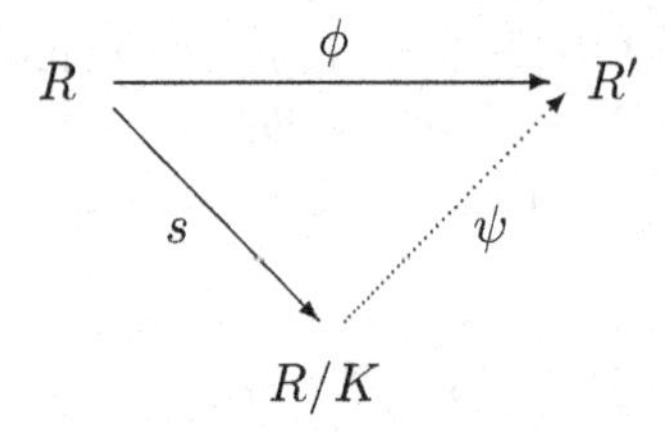

Here is a ring isomorphism based on the Chinese Remainder Theorem.

Proposition 2.41. *Let R be a commutative ring with unity and let N_1, N_2 be ideals of R with $N_1 + N_2 = R$. Then there is a ring isomorphism*

$$\psi : R/N_1N_2 \to R/N_1 \times R/N_2$$

defined by $\psi(a + N_1N_2) = (a + N_1, a + N_2)$, for all $a \in R$.

Proof. Let $\phi : R \to R/N_1 \times R/N_2$ be the map defined by $\phi(a) = (a + N_1, a + N_2)$, for all $a \in R$. For $a, b \in R$,

$$\begin{aligned}\phi(a+b) &= (a+b+N_1, a+b+N_2)\\ &= ((a+N_1)+(b+N_1), (a+N_2)+(b+N_2))\\ &= (a+N_1, a+N_2)+(b+N_1, b+N_2)\\ &= \phi(a)+\phi(b),\end{aligned}$$

$$\begin{aligned}\phi(ab) &= (ab+N_1, ab+N_2)\\ &= ((a+N_1)(b+N_1), (a+N_2)(b+N_2))\\ &= (a+N_1, a+N_2)\cdot(b+N_1, b+N_2)\\ &= \phi(a)\phi(b),\end{aligned}$$

and

$$\phi(1_R) = (1_R+N_1, 1_R+N_2) = 1_{R/N_1\times R/N_2}.$$

Thus ϕ is a homomorphism of rings with unity. Note that $N_1N_2 \subseteq \ker(\phi)$, and so by the UMPK, there exists a homomorphism of rings with unity,

$$\psi : R/N_1N_2 \to R/N_1 \times R/N_2$$

defined by $\psi(a+N_1N_2) = \phi(a) = (a+N_1, a+N_2)$. Let $(a_1+N_1, a_2+N_2) \in R/N_1 \times R/N_2$. By the Chinese Remainder Theorem for Rings, there exists a unique element $a+N_1N_2 \in R/N_1N_2$ for which $\phi(a+N_1N_2) = (a_1+N_1, a_2+N_2)$. It follows that ψ is a ring isomorphism. □

Corollary 2.5. *Let m, n be integers with $\gcd(m,n) = 1$. Then the map $\psi : Z/mnZ \to Z/mZ \times Z/nZ$ defined as $\psi(a+mnZ) = (a+mZ, a+nZ)$ is an isomorphism of rings (cf. Proposition 1.36).*

Let p, q be distinct primes. From Corollary 2.5 one obtains the ring isomorphism $\psi : Z_{pq} \to Z_p \times Z_q$, defined as $a \bmod pq \mapsto (a \bmod p, a \bmod q)$. Consequently,

$$|U(Z_{pq})| = |U(Z_p \times Z_q)| = |U(Z_p)| \cdot |U(Z_q)|,$$

thus yielding the value of Euler's function at pq: $\varphi(pq) = (p-1)(q-1)$.

Proposition 2.42. *Let R be a ring with unity, 1_R. Then the map $\varrho : Z \to R$ defined as $\varrho(n) = n1_R$, where $n1_R$ is defined as in §1.4 is a homomorphism of rings with unity.*

Proof. For $m, n \in Z$,

$$\begin{aligned} \varrho(m+n) &= (m+n)1_R \\ &= m1_R + n1_R \\ &= \varrho(m) + \varrho(n), \end{aligned}$$

and

$$\begin{aligned} \varrho(mn) &= mn1_R \\ &= m(\varrho(n)) \\ &= m(1_R\varrho(n)) \\ &= (m1_R)\varrho(n) \quad \text{by Def. 2.1(iii)} \\ &= \varrho(m)\varrho(n). \end{aligned}$$

Also, $\varrho(1_Z) = 1_Z 1_R = 1_R$, and so, ϱ is a homomorphism of rings with unity. □

The kernel of $\varrho : Z \to R$ is an ideal of Z of the form rZ for some integer $r \geq 0$. The integer r is the **characteristic** of the ring R and is denoted as char(R).

Corollary 2.6. *Let R be a ring with unity with r = char(R). If $r = 0$, then R contains a subring isomorphic to Z. If $r > 0$, then R contains a subring isomorphic to Z_r (a subring is a ring contained in a larger ring).*

Proof. Note that $\varrho(Z)$ is a subring of R. If $r = \text{char}(R) = 0$, then $Z \cong Z/\{0\}$ is isomorphic to $\varrho(Z)$ by Proposition 2.39. If $r = \text{char}(R) > 0$, then $Z_r \cong Z/rZ$ is isomorphic to $\varrho(Z)$ (again by Proposition 2.39). □

2.6 Localization

In this section we continue the main theme of the chapter – the construction of new rings from existing rings. Given a commutative ring with unity R and a multiplicative set S, we construct a ring of fractions with denominators from S that we call the *localization of R at S* and denote by $S^{-1}R$. The contruction is a broad generalization of the construction of the field of rational numbers $\mathbb{Q}$ from the ring of the integers Z.

* * *

Let R be a commutative ring with unity.

Definition 2.22. A subset $S \subseteq R$ is **multiplicatively closed (multiplicative)** if $1 \in S$ and $ab \in S$ whenever $a, b \in S$.

Let $S \subseteq R$ be a multiplicative subset of R. On the cartesian product

$$R \times S = \{(a, b) : a \in R, b \in S\}$$

we define an equivalence relation $\sim$ by the rule $(a, b) \sim (c, d)$ if and only if there exists an element $s \in S$ for which $s(ad - bc) = 0$. Then $\sim$ determines a partition of $R \times S$ into equivalence classes. The collection of all equivalence classes of $\sim$ is denoted by $S^{-1}R$; we let $\frac{a}{b}$ denote the equivalence class containing (a, b).

Proposition 2.43. *Let R be a commutative ring with unity and let $S \subseteq R$ be a multiplicative subset of R. Then $S^{-1}R$ is a commutative ring with unity with addition defined as*

$$\frac{a}{b} + \frac{c}{d} = \frac{ad + bc}{bd}$$

and multiplication defined as

$$\frac{a}{b} \cdot \frac{c}{d} = \frac{ac}{bd},$$

for $\frac{a}{b}, \frac{c}{d} \in S^{-1}R$.

Proof. We first need to show that these relations,

$$+ : S^{-1}R \times S^{-1}R \to S^{-1}R,$$

$$\cdot : S^{-1}R \times S^{-1}R \to S^{-1}R,$$

are actually binary operations on $S^{-1}R$. This amounts to showing that $+$, $\cdot$ are well-defined on equivalence classes, that is, if $(a', b') \sim (a, b)$ and $(c', d') \sim (c, d)$, then $(a'd' + b'c', b'd') \sim (ad + bc, bd)$ and $(a'c', b'd') \sim (ac, bd)$. We leave these straightforward (yet tedious) computations as an exercise.

Now with $0_{S^{-1}R} = \frac{0}{1}$, $\langle S^{-1}R, + \rangle$ is easily shown to be an abelian group. Likewise, the other ring axioms are quickly shown to hold. Finally, $S^{-1}R$ is a commutative ring with unity $1_{S^{-1}R} = \frac{1}{1}$. □

Definition 2.23. The ring $S^{-1}R$ given in Proposition 2.43 is the **localization of R at S.**

We easily see that if $0 \in S$, then $S^{-1}R = \{0\}$.

There is a ring homomorphism $\lambda : R \to S^{-1}R$ defined as $\lambda(a) = a/1$ for $a \in R$. This homomorphism need not be an injection, as we shall see.

Proposition 2.44. *Let $S^{-1}R$ be the localization of R at S and let $\lambda : R \to S^{-1}R$, $a \mapsto a/1$ be the associated ring homomorphism.*

(i) λ is injective if and only if S contains no zero divisors,

(ii) λ is bijective if and only if S consists of units of R.

Proof. For (i) it suffices to show that

$$\ker(\lambda) = \{a \in R : a/1 = 0\} = \{a \in R : sa = 0, s \in S\}.$$

But this follows since $a/1 = 0/1 = 0$ if and only if $s(a \cdot 1 - 1 \cdot 0) = sa = 0$ for some $s \in S$.

For (ii): suppose λ is a bijection. Since λ is an injection, S contains no zero divisors, by (i). Let $b \in S$. Since λ is a surjection, there exists an element $a \in R$ with $\lambda(a) = a/1 = 1/b$, thus $s(ab - 1) = 0$ for some $s \in S$. Hence, b is a unit of R. We leave the converse to the reader. □

The ring homomorphism $\lambda : R \to S^{-1}R$ has the following universal mapping property.

Proposition 2.45 (Universal Mapping Property for Localizations). (UMPL) *Let R be a commutative ring with unity, let S be a multiplicative subset of R, let $S^{-1}R$ be the localization of R at S, and let $\lambda : R \to S^{-1}R$ be the associated ring homomorphism. Let $\phi : R \to R'$ be a homomorphism of commutative rings with unity for which $\phi(S) \subseteq U(R')$. Then there exists a unique ring homomorphism $\psi : S^{-1}R \to R'$ for which the following diagram commutes*

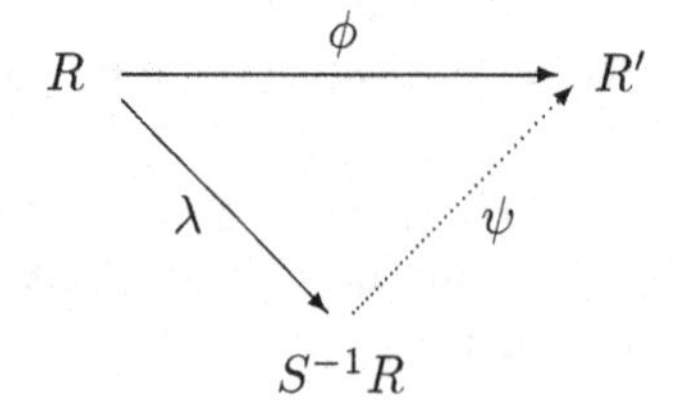

Proof. Define $\psi : S^{-1}R \to R'$ by the rule $\psi(a/b) = \phi(a)\phi(b)^{-1}$. Then as one can check, ψ is well defined on equivalence classes in $S^{-1}R$, and hence gives the required map. $\square$

So to construct a localization, we need a multiplicative set. Here is one way to obtain a multiplicative set. An element $a \in R$ is **nilpotent** if $a^n = 0$ for some $n \geq 1$. For instance, 0 is nilpotent in the ring R and 2 is a nilpotent element of Z_8. Let f be an element of R which is not nilpotent. Then $S = \{1, f, f^2, \dots\}$ is a multiplicative set. The resulting localization $S^{-1}R$ is denoted as R_f.

Note that 3 is a non-nilpotent element of Z_6. Let us construct the localization of Z_6 at $S = \{3^n : \ n \geq 0\} = \{1, 3\}$ (in our notation: this is $(Z_6)_3$). One has

$$Z_6 \times \{1,3\} = \{(0,1),(1,1),(2,1),(3,1),(4,1),(5,1)$$

$$(0,3),(1,3),(2,3),(3,3),(4,3),(5,3)\},$$

and considering the equivalence relation $\sim$ one easily obtains the classes in $(Z_6)_3$:

$$\left\{\frac{0}{1}, \frac{1}{1}\right\}$$

(for instance, $(1,1) \sim (1,3) \sim (3,1) \sim (3,3) \sim (5,1) \sim (5,3)$). Thus $(Z_6)_3 \cong Z_2$. Observe that $\lambda : Z_6 \to (Z_6)_3$ is not injective since 3 is a zero divisor in S (see Proposition 2.44(i)).

Prime ideals of R also provide multiplicative sets. For any sets A, B, let $A \backslash B$ denote the subset of A consisting of all elements of A that are not in B.

Proposition 2.46. *Let N be a prime ideal of R. Then $S = R\backslash N$ is a multiplicative set.*

Proof. Certainly $1 \notin N$ since N is proper. Since N is prime, $ab \in N$ implies that either $a \in N$ or $b \in N$. Thus $a \notin N$ and $b \notin N$ implies that $ab \notin N$. $\square$

By Proposition 2.46, for N prime we obtain the localization $S^{-1}R$, $S = R\backslash N$, which we denote as R_N. If R is an integral domain, then (0) is a prime ideal, and we can form the localization $R_{(0)}$. It is not difficult to show that $R_{(0)}$ is a field, called the **field of fractions** of R and denoted as $\text{Frac}(R)$. Observe that $Z_{(0)} = \text{Frac}(Z) = \mathbb{Q}$. More generally, if F is any field then the ring of polynomials in n variables, $F[x_1, \ldots, x_n]$, is an integral domain and $\mathrm{Frac}(F[x_1, \ldots, x_n])$ is the field of rational functions, denoted as $F(x_1, \ldots, x_n)$.

For another example, let (p) be a non-zero prime ideal of Z. Then $Z\backslash(p)$ is a multiplicative set. The resulting localization $S^{-1}Z = Z_{(p)}$ consists of fractions of the form $\frac{a}{b}$ with $a, b \in Z$, $b \notin (p)$; $A = Z_{(p)}$ is an integral domain with $A_{(0)} = \mathrm{Frac}(A) = \mathbb{Q}$.

Proposition 2.47. *Let N be a prime ideal of R. Then the localization R_N is a local ring with maximal ideal $NR_N = \{r/s : r \in N, s \notin N\}$.*

Proof. Let I be an ideal of R_N, and suppose that I contains an element $x = r/s$ with $r \notin N$. Then $s/r \in R_N$, so that x is a unit of R_N. It follows that $I = R_N$. Thus every proper ideal $I \subseteq R_N$ is contained in NR_N. Consequently, NR_N is a maximal ideal of R_N. If Q is some other maximal ideal of R_N, then $Q \subseteq NR_N$, and so $Q = NR_N$. □

By Proposition 2.47, $Z_{(p)}$ is a local ring with unique maximal ideal $(p) = pZ_{(p)}$.

Localization applied to an integral domain R constructs not only the field of fractions K but also provides a large set of local integral domains R_N between R and K. Often when trying to understand phenomena (such as modules) over R which are understood over K (because modules over a field are vector spaces), it is very helpful to first try to see what's going on over R_N for a non-zero prime ideal of R. (See, for example, the discussion after the proof of Proposition 3.20.) Moreover, understanding the situation over R_N is often facilitated by passing to the completion of R_N, a subject that we take up in the final section of the chapter.

2.7 Absolute Values and Completions

In this final section of the chapter we introduce the notion of distance in $\mathbb{Q}$; we consider the familiar absolute value $|\ |$ on $\mathbb{Q}$ as well as the p-adic absolute value $|\ |_p$ for a prime p. For each absolute value we construct an extension field of $\mathbb{Q}$ called the completion of $\mathbb{Q}$. The completion of $\mathbb{Q}$ with respect to $|\ |$ is the field of real numbers $\mathbb{R}$, the completion with respect to $|\ |_p$ is the field of p-adic rationals $\mathbb{Q}_p$. The valuation ring of $\mathbb{Q}_p$ is the ring of p-adic integers $\mathbb{Z}_p$ which is a ring extension of Z.

* * *

Definition 2.24. An **absolute value** on $\mathbb{Q}$ is a function $|\ |: \mathbb{Q} \to \mathbb{R}$ which satisfies for all $x, y \in \mathbb{Q}$,

(i) $|x| \geq 0$,

(ii) $|x| = 0$ if and only if $x = 0$,

(iii) $|xy| = |x||y|$,

(iv) $|x + y| \leq |x| + |y|$.

Consider the rationals $\mathbb{Q}$ with the ordinary absolute value $|\ |$. A sequence $\{a_n\}$ in $\mathbb{Q}$ is a $|\ |$**-Cauchy sequence** if for each $\iota > 0$ there exists an integer N for which $|a_m - a_n| < \iota$ whenever $m, n \geq N$. For instance, the sequence

$$1, \frac{1}{2}, \frac{1}{3}, \frac{1}{4}, \ldots$$

is a $|\ |$-Cauchy sequence in $\mathbb{Q}$. This sequence converges to a limiting value of 0, which is an element of $\mathbb{Q}$. For another example, consider the sequence

$$1, 1.4, 1.41, 1.414, 1.4142, 1.41421, 1.414213, 1.4142135, 1.41421356, \ldots$$

where a_n, $n \geq 1$, is the first n digits of the decimal expansion of $\sqrt{2}$. As one can verify, this is a $|\ |$-Cauchy sequence in $\mathbb{Q}$. In this case, the limit of the sequence is $\sqrt{2}$ which is not an element of $\mathbb{Q}$. We say that $\mathbb{Q}$ is not complete with respect to the absolute value $|\ |$ since there exists a $|\ |$-Cauchy sequence in $\mathbb{Q}$ that does not converge to an element in $\mathbb{Q}$.

There exists a field extension K of $\mathbb{Q}$ with the following properties:

(i) The absolute value $|\ |$ extends uniquely to an absolute value on K, also denoted as $|\ |$,

(ii) with respect to $|\ |$, $\mathbb{Q}$ is dense in K, that is, the closure $\overline{\mathbb{Q}} = K$,

(iii) K is complete with respect to $|\ |$, that is, every $|\ |$-Cauchy sequence in K converges to an element in K.

This field extension K is (of course!) the field of real numbers $\mathbb{R}$, and since $\mathbb{R}$ satisfies conditions (i), (ii), (iii), we say that $\mathbb{R}$ is the **completion of $\mathbb{Q}$ with respect to $|\ |$**.

We can define another kind of absolute value on $\mathbb{Q}$. Let $x \in \mathbb{Q}$ and let p be a prime number. The **p-adic absolute value** $|\ |_p$ is defined as $|x|_p = 0$ if $x = 0$ and

$$|x|_p = \frac{1}{p^r}$$

if $x = p^r(s/t) \in \mathbb{Q}$ with $\gcd(s,p) = 1$, $\gcd(t,p) = 1$, $r \in Z$. For instance, $|15|_3 = \frac{1}{3}$, and $|\frac{2}{9}|_3 = 9$. One should indeed check that $|\ |_p$ defines an absolute value on $\mathbb{Q}$ (see §2.8, Exercise 90). Note that $|\ |_p$ satisfies a stronger condition than (iv) of Definition 2.24.

Proposition 2.48. *Let p be a prime number and let $|\ |_p$ denote the p-adic absolute value on $\mathbb{Q}$. Then for all $x, y \in \mathbb{Q}$,*

(i) $|x+y|_p \le max\{|x|_p, |y|_p\}$,

(ii) If $|x|_p \ne |y|_p$, then $|x+y|_p = max\{|x|_p, |y|_p\}$.

Proof. Exercise. □

A sequence $\{a_n\}$ in $\mathbb{Q}$ is a $|\ |_p$-**Cauchy sequence** if for each $\iota > 0$ there exists an integer N for which $|a_m - a_n|_p < \iota$ whenever $m, n \ge N$. For example, the sequence

$$1, p, p^2, p^3, p^4, \ldots \tag{2.2}$$

is a $|\ |_p$-Cauchy sequence in $\mathbb{Q}$; the sequence

$$1, \frac{1}{p}, \frac{1}{p^2}, \frac{1}{p^3}, \frac{1}{p^4}, \ldots$$

however, is not a $|\ |_p$-Cauchy sequence!

The $|\ |_p$-Cauchy sequence (2.2) converges to the element $0 \in \mathbb{Q}$. As in the case where $\mathbb{Q}$ is endowed with the ordinary absolute value, not every $|\ |_p$-Cauchy sequence in $\mathbb{Q}$ converges to an element of $\mathbb{Q}$. In other words, the field $\mathbb{Q}$ is not complete with respect to the p-adic absolute value.

Proposition 2.49. *For each prime p there exists a $|\ |_p$-Cauchy sequence in $\mathbb{Q}$ that does not converge to an element of $\mathbb{Q}$.*

Proof. The prime $p = 2$ is special, and so we will handle it first. We construct an explicit example of a $|\ |_2$-Cauchy sequence in $\mathbb{Q}$ that does not converge to an element of $\mathbb{Q}$. Consider $x^3 - 3$, which is irreducible over $\mathbb{Q}$. The conguence $x^3 \equiv 3 \bmod 2$ has solution $x_1 = 1$; $x_1 = 1$ is the first term in our Cauchy sequence. To compute the second term x_2, we find an integer $k_1 \ge 0$ so that $(1+2k_1)^3 \equiv 3 \bmod 4$. Thus:

$$\begin{aligned}
(1+2k_1)^3 &\equiv 3 \bmod 4\\
1 + 6k_1 + 12k_1^2 + 8k^3 &\equiv 3 \bmod 4\\
2 + 2k_1 &\equiv 0 \bmod 4\\
1 + k_1 &\equiv 0 \bmod 2,
\end{aligned}$$

so that $k_1 = 1$. Now let $x_2 = 1 + 2k_1 = 1 + 2 \cdot 1 = 3$.

To obtain x_3 we find an integer $k_2 \geq 0$ so that $(3 + 4k_2)^3 \equiv 3 \bmod 8$:

$$\begin{aligned}(3 + 4k_2)^3 &\equiv 3 \bmod 8\\ 27 + 108k_2 + 144k_2^2 + 64k_2^3 &\equiv 3 \bmod 8\\ 24 + 108k_2 &\equiv 0 \bmod 8\\ 6 + 27k_2 &\equiv 0 \bmod 2\\ k_2 &\equiv 0 \bmod 2,\end{aligned}$$

so that $k_2 = 0$. Now, let $x_3 = 3 + 4k_2 = 3 + 4 \cdot 0 = 3$.

To obtain x_4, we find $k_3 \geq 0$ so that $(3 + 8k_3)^3 \equiv 3 \bmod 16$:

$$\begin{aligned}(3 + 8k_3)^3 &\equiv 3 \bmod 16\\ 27 + 216k_3 + 576k_3^2 + 512k_3^3 &\equiv 3 \bmod 16\\ 24 + 216k_3 &\equiv 0 \bmod 16\\ 3 + 27k_3 &\equiv 0 \bmod 2\\ k_3 &\equiv 1 \bmod 2,\end{aligned}$$

so that $k_3 = 1$. Now, let $x_4 = 3 + 8k_3 = 3 + 8 \cdot 1 = 11$.

Continuing in this manner, we obtain the $|\ |_2$-Cauchy sequence

$$1, 3, 3, 11, \ldots$$

whose limiting value is a root of $x^3 - 3$, and hence cannot be an element of $\mathbb{Q}$.

Next we consider the case $p > 2$. Let a be a non-square integer so that the polynomial $x^2 - a \in Z[x]$ is irreducible over $\mathbb{Q}$. Assume that $a \not\equiv 0 \bmod p$ and that the conguence $x^2 \equiv a \bmod p$ has a solution, which we denote as x_1. Now, there exists an integer $k_1 \geq 0$ so that the conguence $x^2 \equiv a \bmod p^2$ has solution $x_2 = x_1 + pk_1$. Moreover, there exists an integer $k_2 \geq 0$ so that the congruence $x^2 \equiv a \bmod p^3$ has solution $x_3 = x_2 + p^2k_2$. Continuing in this manner, we obtain a $|\ |_p$-Cauchy sequence $\{x_n\}$ whose limiting value is a root of $x^2 - a$ and hence cannot be an element of $\mathbb{Q}$. Try it! (See §2.8, Exercise 94). □

Analogous to the completion $\mathbb{R}$ of $\mathbb{Q}$, we seek to construct a field extension K of $\mathbb{Q}$ with the following properties:

(i) The p-adic absolute value $|\ |_p$ extends uniquely to an absolute value on K, also denoted as $|\ |_p$,

(ii) with respect to $|\ |_p$, $\mathbb{Q}$ is dense in K, that is, the closure $\overline{\mathbb{Q}} = K$,

(iii) K is complete with respect to $| \ |_p$, that is, every $| \ |_p$-Cauchy sequence in K converges to an element in K.

A field extension K satisfying these conditions is the **completion of $\mathbb{Q}$ with respect to $| \ |_p$**. As we shall see, the completion of $\mathbb{Q}$ with respect to $| \ |_p$ is the **field of p-adic rationals**, denoted by $\mathbb{Q}_p$.

We begin by constructing a certain field extension of $\mathbb{Q}$. Let $\mathcal{C}$ denote the set of all $| \ |_p$-Cauchy sequences in $\mathbb{Q}$. Let $\{x_n\}$, $\{y_n\} \in \mathcal{C}$. The sum and product of the sequences are defined termwise as

$$\{x_n\} + \{y_n\} = \{x_n + y_n\},$$

$$\{x_n\} \cdot \{y_n\} = \{x_n y_n\}.$$

Proposition 2.50. *Let $\{x_n\}$, $\{y_n\} \in \mathcal{C}$. Then $\{x_n\} + \{y_n\}$ and $\{x_n\} \cdot \{y_n\}$ are Cauchy sequences.*

Proof. Let $\iota > 0$. Since $\{x_n\}$ and $\{y_n\}$ are Cauchy sequences, there exists integers N_1, N_2 so that $m, n \geq N_1$ implies that $|x_m - x_n|_p < \iota$ and $m, n \geq N_2$ implies $|y_m - y_n|_p < \iota$. Let $N = \max\{N_1, N_2\}$. Then

$$\begin{aligned} |(x_m + y_m) - (x_n + y_n)|_p &= |x_m - x_n + y_m - y_n|_p \\ &\leq \max\{|x_m - x_n|_p, |y_m - y_n|_p\} \\ &< \iota, \end{aligned}$$

whenever $m, n \geq N$. Thus $\{x_n\} + \{y_n\} \in \mathcal{C}$. We leave it as an exercise to show that $\{x_n\} \cdot \{y_n\} \in \mathcal{C}$. $\square$

The binary operations $+, \cdot$ endow $\mathcal{C}$ with the structure of a commutative ring with unity. For $0_{\mathcal{C}}$, we take the constant sequence $\{0\}$, and for $1_{\mathcal{C}}$, we take the constant sequence $\{1\}$.

We identify a special subset of $\mathcal{C}$, namely those Cauchy sequences $\{x_n\}$ that converge to 0 in the usual sense: given $\iota > 0$ there exists an integer N so that

$$|x_n - 0|_p = |x_n|_p < \iota, \quad \text{whenever } n \geq N.$$

For these sequences, we write $\lim_{n \to \infty} x_n = 0$. Let

$$\mathcal{N} = \{\{x_n\} \in \mathcal{C} : \lim_{n \to \infty} x_n = 0\}.$$

Lemma 2.1. *Let $\{x_n\} \in \mathcal{C} \backslash \mathcal{N}$. Then there exists $c > 0$ and an integer N so that $|x_n|_p \geq c$ for all $n \geq N$.*

Proof. If no such c and N exist, then for each $k \geq 1$, there exists a term x_{n_k} of $\{x_n\}$ with $|x_{n_k}|_p < \frac{1}{k}$. Thus $\{x_n\}$ contains a subsequence $\{x_{n_k}\}$ with $\lim_{k\to\infty} x_{n_k} = 0$. But since $\{x_n\}$ is a Cauchy sequence, $\{x_n\}$ converges to 0. To see this, let $\iota > 0$. Then there exist integers N_1, N_2 so that $|x_m - x_n|_p < \iota/2$ whenever $m, n \geq N_1$, and $|x_{n_k}|_p < \iota/2$ whenever $k \geq N_2$. Let $M = \max\{N_1, N_2\}$ and note that $n_M \geq M$. Then

$$|x_m|_p = |(x_m - x_{n_k}) + x_{n_k}|_p \leq |x_m - x_{n_k}|_p + |x_{n_k}|_p < \iota$$

whenever $m \geq M$. Thus $\{x_m\}$ converges to 0, in other words, $\{x_m\} \in \mathcal{N}$, a contradiction. □

Proposition 2.51. *$\mathcal{N}$ is a maximal ideal of $\mathcal{C}$.*

Proof. It is an easy exercise (see §2.8, Exercise 95) to show that $\mathcal{N}$ is an ideal of $\mathcal{C}$.

To prove that $\mathcal{N}$ is a maximal ideal we proceed as follows. Certainly $\mathcal{N}$ is a proper ideal since the constant sequence $\{x\}$, $x \in \mathbb{Q}^\times$ does not converge to 0. Suppose that $\mathcal{I}$ is an ideal of $\mathcal{C}$ with

$$\mathcal{N} \subset \mathcal{I} \subseteq \mathcal{C}.$$

Let $\{x_n\} \in \mathcal{I}\backslash\mathcal{N}$ and let $\mathcal{J}$ be the ideal of $\mathcal{C}$ generated by $\{\{x_n\}\} \cup \mathcal{N}$. Then $\mathcal{J} \subseteq \mathcal{I}$. We show that $\mathcal{J} = \mathcal{C}$, which yields $\mathcal{I} = \mathcal{C}$. To this end, by Lemma 2.1, there exists $c > 0$ and an integer N so that $|x_n|_p \geq c$ for all $n \geq N$. Define a sequence $\{y_n\}$ by setting $y_n = 0$ for $n < N$ and $y_n = \frac{1}{x_n}$ for $n \geq N$. Then $\{y_n\}$ is a Cauchy sequence. Here is the proof.

Let $\iota > 0$. There exist N' so that for $m, n \geq N'$, $|x_m - x_n|_p < c^2\iota$. Let $M = \max\{N, N'\}$. Now for $m, n \geq M$,

$$\begin{aligned}
\iota &> \frac{|x_m - x_n|_p}{c^2} \\
&\geq \frac{|x_m - x_n|_p}{|x_m x_n|_p} \\
&= \left|\frac{1}{x_m} - \frac{1}{x_n}\right|_p \\
&= |y_m - y_n|_p.
\end{aligned}$$

And so, $\{y_n\}$ is a Cauchy sequence.

Now, the sequence $\{x_n\} \cdot \{y_n\}$ has terms that are 0 for $n < N$ and 1 for $n \geq N$. Hence the sequence $\{1\} - \{x_n\} \cdot \{y_n\} \in \mathcal{N}$. Consequently, $\{1\} \in \mathcal{J}$, whence $\mathcal{J} = \mathcal{C}$. □

By Proposition 2.34, $\mathcal{C}/\mathcal{N}$ is a field. The field $\mathcal{C}/\mathcal{N}$ is the field of **p-adic rational numbers** denoted as $\mathbb{Q}_p$. We identify an element $x \in \mathbb{Q}$ with the coset $\{x\} + \mathcal{N}$ represented by the constant sequence $\{x\}$, and so $\mathbb{Q}_p$ is a field extension of $\mathbb{Q}$. Our goal is to show that $\mathbb{Q}_p$ is the completion of $\mathbb{Q}$ with respect to $| \ |_p$.

We show that the absolute value $| \ |_p$ extends to an absolute value on $\mathbb{Q}_p$.

Proposition 2.52. *The p-adic absolute value $| \ |_p$ on $\mathbb{Q}$ extends uniquely to an absolute value on $\mathbb{Q}_p$.*

Proof. Let λ be an element of $\mathbb{Q}_p$ represented by the Cauchy sequence $\{x_n\}$ (that is, $\lambda = \{x_n\} + \mathcal{N}$). If $\{x_n\} \in \mathcal{N}$, then we set $|\lambda|_p = 0$. Else, assume that $\{x_n\} \in \mathcal{C}\backslash\mathcal{N}$. By Lemma 2.1, there exist $c > 0$ and an integer N_1 so that $|x_n|_p \geq c$ for all $n \geq N_1$. But since $\{x_n\}$ is Cauchy, there exists an integer N_2 so that $|x_m - x_n|_p < c$ whenever $m, n \geq N_2$. Let $N = \max\{N_1, N_2\}$. Then for $m, n \geq N$,

$$\max\{|x_m|_p, |x_n|_p\} \geq c > |x_m - x_n|_p.$$

If $|x_m|_p \neq |x_n|_p$ for any $m, n \geq N$, then

$$|x_m - x_n|_p = \max\{|x_m|_p, |x_n|_p\}$$

by Proposition 2.48(ii), and so,

$$\max\{|x_m|_p, |x_n|_p\} > |x_m - x_n|_p = \max\{|x_m|_p, |x_n|_p\},$$

a contradiction. Consequently, $|x_m|_p = |x_n|_p$ for all $m, n \geq N$. It follows that $\lim_{n\to\infty} |x_n|_p$ exists. We define

$$|\lambda|_p = \lim_{n\to\infty} |x_n|_p.$$

In this manner, $\mathbb{Q}_p$ is endowed with an absolute value which is the unique extension of $| \ |_p$ to $\mathbb{Q}_p$, we denote this extension by $| \ |_p$. □

We next show that $\mathbb{Q}$ is dense in $\mathbb{Q}_p$.

Proposition 2.53. *The field $\mathbb{Q}$ is a dense subset of $\mathbb{Q}_p$, that is, given $\lambda \in \mathbb{Q}_p$ and $\iota > 0$, there exists an $y \in \mathbb{Q}$ for which $|y - \lambda|_p < \iota$.*

Proof. Recall that λ is really a left coset, so let $\{x_n\}$ represent λ. Let $\iota > 0$. Now, since $\{x_n\}$ is Cauchy, there exists N so that $m, n \geq N$ implies $|x_m - x_n|_p < \iota/2$. Let $\{y\}$ be the constant sequence with $y = x_N$. Then for $n \geq N$, $|x_N - x_n|_p < \iota/2$. Thus $\lim_{n\to\infty} |x_N - x_n|_p \leq \iota/2 < \iota$. Now considering the sequence $\{x_N - x_n\}$ as a representative of the element $y - \lambda \in \mathbb{Q}_p$, one has $|y - \lambda|_p < \iota$. Thus $\mathbb{Q}$ is dense in $\mathbb{Q}_p$. □

Finally, to prove that $\mathbb{Q}_p$ is the completion of $\mathbb{Q}$, we show that $\mathbb{Q}_p$ is complete with respect to the absolute value $| \ |_p$.

Proposition 2.54. *The field of p-adic rationals $\mathbb{Q}_p$ is complete with respect to the extended p-adic absolute value $| \ |_p$.*

Proof. Let $\{\lambda_n\}$ be a $| \ |_p$-Cauchy sequence in $\mathbb{Q}_p$. (Of course, $\{\lambda_n\}$ is really a sequence of cosets in $\mathcal{C}/\mathcal{N}$ represented by a sequence of Cauchy sequences in $\mathbb{Q}$, $\lambda_1, \lambda_2, \ldots$, which satisfies the condition, for $\iota > 0$, there exists an integer N fo which $|\lambda_m - \lambda_n|_p < \iota$ whenever $m, n \geq N$. Here $| \ |_p$ is the extended absolute value.)

By Proposition 2.53, for each $n \geq 1$, there exists a constant sequence $y^{(n)} \in \mathbb{Q}$ for which

$$|y^{(n)} - \lambda_n|_p < 10^{-n},$$

and so,

$$\lim_{n\to\infty} |y^{(n)} - \lambda_n|_p = 0.$$

Now, for $\iota > 0$, there exists an integer N_1 so that

$$|\lambda_n - y^{(n)}|_p < \iota/3,$$

whenever $n \geq N_1$. Since $\{\lambda_n\}$ is Cauchy, there exists an integer N_2 so that

$$|\lambda_m - \lambda_n|_p < \iota/3$$

whenever $m, n \geq N_2$. Take $N = \max\{N_1, N_2\}$. Thus,

$$|\lambda_m - y^{(n)}|_p < 2\iota/3,$$

whenever $m, n \geq N$. Observe that

$$|y^{(m)} - \lambda_m|_p < \iota/3$$

since $m \geq N \geq N_1$, hence

$$|y^{(m)} - y^{(n)}|_p < \iota.$$

It follows that

$$|x^{(m)} - x^{(n)}|_p < \iota$$

whenever $m, n \geq N$, where $x^{(n)} \in \mathbb{Q}$ is the constant term of $y^{(n)}$. Thus $\{x^{(n)}\}$ is a Cauchy sequence in $\mathbb{Q}$ which converges to the coset $\lambda = \{x^{(n)}\} + \mathcal{N} \in \mathbb{Q}_p$. Thus

$$\lim_{n\to\infty} y^{(n)} = \lambda.$$

Now there exists an integer N_3 so that

$$|y^{(n)} - \lambda|_p < 2\iota/3$$

whenever $n \geq N_3$. Let $M = \max\{N_1, N_3\}$. Then

$$|\lambda_n - \lambda|_p < \iota,$$

whenever $n \geq M$, so that

$$\lim_{n\to\infty} \lambda_n = \lambda,$$

thus $\mathbb{Q}_p$ is complete. □

There is a subring of $\mathbb{Q}_p$ defined as

$$\{x \in \mathbb{Q}_p : |x|_p \leq 1\}.$$

This is the **ring of p-adic integers**, and is denoted by $\mathbb{Z}_p$.

Proposition 2.55. *Every (coset) $x \in \mathbb{Z}_p$ has a representative of the form $\{t_n\}$, with*

$$t_1 = a_0,$$

$$t_2 = a_0 + a_1 p,$$

$$t_3 = a_0 + a_1 p + a_2 p^2,$$

$$\vdots$$

$$t_n = a_0 + a_1 p + a_2 p^2 + a_3 p^2 + \cdots + a_{n-1} p^{n-1},$$

with $0 \leq a_i \leq p-1$. We write the coset $\{t_n\} + \mathcal{N}$ as the infinite sum

$$a_0 + a_1 p + a_2 p^2 + a_3 p^3 + \cdots,$$

thus

$$x = a_0 + a_1 p + a_2 p^2 + a_3 p^3 + \cdots.$$

Proof. *Step 1.* Let $n \geq 1$. By Proposition 2.53 there exists an element $\frac{p^r a}{b} \in \mathbb{Q}$, $\gcd(a,p) = \gcd(b,p) = 1$, $r \in Z$, so that

$$\left| \frac{p^r a}{b} - x \right|_p \leq \frac{1}{p^n} < 1. \tag{2.3}$$

Consequently,

$$\left| \frac{p^r a}{b} \right|_p \leq \max \left\{ |x|_p, \left| \frac{p^r a}{b} - x \right|_p \right\} \leq 1,$$

and so, $r \geq 0$. Since $\gcd(b,p) = 1$ there exists an integer b' with $bb' \equiv 1 \bmod p^n$. Thus

$$|p^r abb' - p^r a|_p \leq \frac{1}{p^n}$$

and so

$$\left| p^r ab' - \frac{p^r a}{b} \right|_p \leq \frac{1}{p^n},$$

since $|b|_p = 1$. Now from (2.3) one obtains

$$|p^r ab' - x|_p \leq \frac{1}{p^n}.$$

Taking $t = p^r ab' \bmod p^n$, one sees that there exists an integer t, $0 \leq t \leq p^n - 1$, so that

$$|t - x|_p \leq \frac{1}{p^n}.$$

Step 2. Applying Step 1 to each integer $n \geq 1$, construct a sequence of integers $\{t_n\}$ with the property that

$$|t_n - x|_p \leq \frac{1}{p^n},$$

$0 \leq t_n \leq p^n - 1$. Now, $\{t_n\}$ represents x. Observe that

$$|t_{n+1} - t_n|_p \leq \frac{1}{p^n}, \tag{2.4}$$

for $n \geq 1$. Put $a_0 = t_1$ and write t_2 in powers of p as $t_2 = b_0 + b_1 p$, $0 \leq b_0, b_1 \leq p - 1$. Then from (2.4) with $n = 1$, one has $b_0 = a_0$. Let $a_1 = b_1$. Now write $t_3 = c_0 + c_1 p + c_2 p^2$ with $0 \leq c_0, c_1, c_2 \leq p - 1$. Then from (2.4) with $n = 2$ one has $c_0 = a_0$, $c_1 = a_1$. Let $a_2 = c_2$. Continuing in this manner, one has for $n \geq 1$,

$$t_n = a_0 + a_1 p + a_2 p^2 + a_3 p^3 + \cdots + a_{n-1} p^{n-1}$$

with $0 \leq a_i \leq p - 1$. It follows that $x \in \mathbb{Z}_p$ has a representative of the claimed form. □

Proposition 2.56. *Let $\mathbb{Z}_p$ denote the ring of p-adic integers. Then the following hold.*

(i) The group of units of $\mathbb{Z}_p$ is given as

$$U(\mathbb{Z}_p) = \{x \in \mathbb{Z}_p : |x|_p = 1\},$$

(ii) $\mathbb{Z}_p$ is a local ring with maximal ideal

$$p\mathbb{Z}_p = \{x \in \mathbb{Z}_p : |x|_p < 1\}.$$

Proof. For (i), Let $x \in \mathbb{Z}_p$ with $|x|_p = 1$. Then the multiplicative inverse x^{-1} satisfies $|x^{-1}|_p = 1/|x|_p = 1$. Thus $x \in U(\mathbb{Z}_p)$. Conversely, suppose that $x \in U(\mathbb{Z}_p)$ with $y = x^{-1}$. Then $1 = xy$, hence $1 = |x|_p|y|_p$. Thus $|x|_p = 1$ since $|x|_p \leq 1$ and $|y|_p \leq 1$.

For (ii): Let

$$x = a_0 + a_1p + a_2p^2 + a_3p^2 + \cdots,$$

$0 \leq a_i \leq p-1$, be an element of $\mathbb{Z}_p$. Then $|px|_p \leq \frac{1}{p} < 1$ as one can easily check. Also, any $x \in \mathbb{Z}_p$ with $|x|_p < 1$ must have a representative of the form

$$x = a_0 + a_1p + a_2p^2 + a_3p^2 + \cdots,$$

with $a_0 = 0$, hence $x \in p\mathbb{Z}_p$. Thus $p\mathbb{Z}_p = \{x \in \mathbb{Z}_p : |x|_p < 1\}$. Now let I be an ideal of $\mathbb{Z}_p$ that contains an element not in $p\mathbb{Z}_p$. Let x be such an element. Then $|x|_p = 1$ hence x is a unit, and thus $I = \mathbb{Z}_p$ by Proposition 2.17. It follows that $p\mathbb{Z}_p$ is the unique maximal ideal of $\mathbb{Z}_p$. □

Proposition 2.57. *Every element of $\mathbb{Q}_p$ has a representative of the form*

$$\sum_{i=-r}^{\infty} a_ip^i,$$

where r is an integer and $0 \leq a_i \leq p-1$ for $i = -r, -r+1, -r+2, \ldots$.

Proof. Let $x \in \mathbb{Q}_p$. If $x = 0$, then $x = 0 + 0 \cdot p + 0 \cdot p^2 + \cdots$. So we assume $x \neq 0$. Then $|x|_p = \frac{1}{p^r}$ for some integer r. Consequently $p^rx \in \mathbb{Z}_p$, and so by Proposition 2.55,

$$p^rx = a_0 + a_1p + a_2p^2 + \cdots,$$

with $0 \leq a_i \leq p-1$. Thus,

$$x = a_0p^{-r} + a_1p^{-r+1} + a_2p^{-r+2} + \cdots,$$

which upon renumbering is of the claimed form. □

Let $x \in \mathbb{Q}_p$ and write

$$x = \sum_{i=-r}^{\infty} a_i p^i,$$

where r is an integer and $0 \leq a_i \leq p-1$ for $i \geq -r$. The smallest integer s for which $a_s \neq 0$ is the **p-order** of x, denoted by $\mathrm{ord}_p(x)$. Note that $\mathrm{ord}_p(x) = s$ if and only if $|x|_p = p^{-s}$. We have $\mathrm{ord}_p(0) = \infty$, which makes sense since $|0|_p = p^{-\infty} = 0$. One has $x \in \mathbb{Z}_p$ if and only if $\mathrm{ord}_p(x) \geq 0$.

Proposition 2.58. *Every $x \in \mathbb{Q}_p$ can be written in the form*

$$x = up^s,$$

where u is a unit of $\mathbb{Z}_p$ and $s = ord_p(x)$.

Proof. If $x = 0$, then $x = 1 \cdot p^{\infty}$, so we assume that $x \neq 0$. By Proposition 2.57, we can write an element $x \in \mathbb{Q}_p$ in the form

$$x = \sum_{i=s}^{\infty} a_i p^s = a_s p^s + a_{s+1} p^{s+1} + a_{s+2} p^{s+2} + \cdots$$

where $s = \mathrm{ord}_p(x)$ is an integer and $0 \leq a_i \leq p-1$ for $i \geq s$. Now,

$$x = p^s \sum_{i=0}^{\infty} a_{s+i} p^i,$$

with $|\sum_{i=0}^{\infty} a_{s+i} p^i|_p = 1$. The result follows. □

Here is an easy corollary that we leave as an exercise.

Corollary 2.7. *$\mathbb{Z}_p$ is an integral domain with $Frac(\mathbb{Z}_p) = \mathbb{Q}_p$.*

Proof. Exercise. □

2.8 Exercises

Exercises for §2.1

(1) Let R be a ring. Prove that $(-a)(-b) = ab$ for all $a, b \in R$.
(2) Let $f(x, y) = 3 + x + xy - x^2y^3 + 6xy^4$. What is the degree of $f(x, y)$ considered as an element of $Z[x][y]$? What is the degree of $f(x, y)$ considered as an element of $Z[y][x]$?
(3) Prove that Z_p is a field for any prime p.
(4) Find all of the units in the ring $Z \times Z_2$.

(5) Determine whether $\begin{pmatrix} 3 & 1 & 2 \\ 1 & 0 & 1 \\ 0 & 1 & 1 \end{pmatrix}$ is a unit in $\mathrm{Mat}_3(\mathbb{R})$. Is this matrix a unit in $\mathrm{Mat}_3(Z)$?

(6) Let R be a commutative ring with unity and let $a, b, c \in R$. Prove the following: if $a \mid b$ and $b \mid c$, then $a \mid c$.

(7) Find a greatest common divisor of $x^2 + 2x - 3$ and $x^5 - 3x^2 - 2x - 4$ in $\mathbb{Q}[x]$.

(8) Let R be a commutative ring with unity. Prove the following: if d is a greatest common divisor of a, b and u is any unit of R, then ud is a greatest common divisor of a, b.

(9) Let R be an integral domain, let $a, b \in R$ and let d be a greatest common divisor of a, b. Prove the following: d' is a greatest common divisor of a, b if and only if $d' = ud$ for some unit $u \in R$.

(10) Let C_4 denote the cyclic group of order 4 generated by g and let ZC_4 be the group ring over Z. Compute a greatest common divisor for $1 - g^2$, $2 + 2g + 2g^2 + 2g^3$.

(11) Let $Z\mathcal{V}$ denote the group ring of the Klein 4-group over Z.

   (a) Compute $(1 + 2a + 3b - c)(a + b + c)$.
   (b) Find a zero divisor in $Z\mathcal{V}$.
   (c) Find all of the units in $Z\mathcal{V}$.

(12) Find all of the zero divisors in ZD_3.

(13) Determine whether there exists a sequence $\{a_n\} \subseteq ZC_3$ with the following properties: (i) $a_{n+1} | a_n$ for all n, (ii) there is no positive integer N for which $a_i = ua_{i+1}$, for some unit u, whenever $i \geq N$.

Exercises for §2.2

(14) Prove the uniqueness part of the Division Theorem.

(15) Let $f(x) \in \mathbb{Q}[x]$. Prove that there is an element $r \in \mathbb{Q}$ so that $rf(x)$ is a primitive polynomial.

(16) Prove that $x^4 + 4x^3 + 6x^2 + 4x - 1$ is irreducible over $\mathbb{Q}$.

(17) Show that $x^4 + 4$ is reducible over $\mathbb{Q}$.

(18) How many zeros does $x^2 - 1$ have in Z_8? Why doesn't this contradict Proposition 2.6?

Exercises for §2.3

(19) If p is a prime number, then the residue class ring Z_p is a field (§2.8,

Exercise 3). Thus by Proposition 2.15, $Z_p^\times = \{1, 2, \ldots, p-1\}$ is cyclic. A residue $m \in Z_p^\times$ for which $\langle m \rangle = Z_p^\times$ is a **primitive root modulo** p.

(a) Show that 2 is a primitive root modulo 5; show that 6 is a primitive root modulo 13.
(b) Find all of the primitive roots modulo 17.

(20) Suppose that $U(Z_n)$ is cyclic. Does it necessarily follow that n is prime?
(21) Let p, q be prime numbers with $\gcd(p-1, q-1) = 1$. Show that either $U(Z_p \times Z_q) \cong Z_{p-1}$ or $U(Z_p \times Z_q) \cong Z_{q-1}$.
(22) Show that $|U(Z_{2p})| = |U(Z_p)|$ for $p > 2$ prime. Is $U(Z_{2p})$ isomorphic to $U(Z_p)$?
(23) Prove Proposition 2.11.
(24) Compute $457^{67} \bmod 23$.
(25) Give the complete factorization of the polynomial $x^6 - 1$ over Z_7.
(26) Find all of the generators for U_5 and U_6.
(27) Find a group which is isomorphic (but not equal) to $U(Z_4 \times Z_4)$.
(28) Compute $U(Z_3C_2)$ and $U(\mathbb{Q}C_2)$.
(29) Prove that the group $U(ZC_3)$ is cyclic.
(30) Let G be the cyclic group of order m. Prove that $\hat{G} \cong G$.
(31) Let ζ_p denote a primitive pth root of unity.

(a) Verify the factorization $x^p - 1 = \prod_{i=0}^{p-1}(x - \zeta_p^i)$ over $\mathbb{C}$.
(b) Use (a) to prove that $p = \prod_{i=1}^{p-1}(1 - \zeta_p^i)$ and $\sum_{i=0}^{p-1} \zeta_p^i = 0$.

(32) Show that $\zeta_p + \zeta_p^{-1}$ is a root of a monic polynomial with coefficients in $\mathbb{Q}$.
(33) Let $p > 3$ be prime and let ζ_6 denote a primitive 6th root of unity. Show that $\zeta_6^{p^2} = \zeta_6$.
(34) Let p be prime. For $2 \le a \le (p-1)/2$, $\gcd(a, p) = 1$, show that

$$\zeta_p^{(1-a)/p}\left(\frac{1 - \zeta_p^a}{1 - \zeta_p}\right)$$

is a real number.
(35) Find all of the characters of D_3.
(36) Compute the character group of C_3 where the characters are homomorphisms $\gamma : C_3 \to \mathbb{R}^\times$.
(37) Compute the character group of D_3 where the characters are homomorphisms $\gamma : D_3 \to \mathbb{R}^\times$.

(38) In the manner of Example 2.9, represent the elements of D_4 as 8 plane isometries of the square.

Exercises for §2.4

(39) Let I, J be ideals of the ring R.

(a) Prove that $I \cap J$ is an ideal.
(b) Prove that $I + J = \{a + b : a \in I, b \in J\}$ is an ideal.
(c) Give a counterexample to show that $I \cup J$ is not an ideal.

(40) Find an example of a ring R and ideals I, J for which $IJ \subset I \cap J$.
(41) Prove that the ideal $(2, x) \subseteq Z[x]$ is maximal.
(42) Prove that the ideal $(x) \subseteq Z[x]$ is prime, but the ideal $(x) \subseteq Z_6[x]$ is not prime.
(43) Find a counterexample which shows that the converse of Proposition 2.22 is false.
(44) Let $a + b\sigma + c\sigma^2 \in ZC_3$, $\langle\sigma\rangle = C_3$.

(a) Verify the formula

$$(a+b\sigma+c\sigma^2)((a^2-bc)+(c^2-ab)\sigma+(b^2-ac)\sigma^2) = a^3+b^3+c^3-3abc.$$

(b) Use (a) to show that $(7, 2 - \sigma)$ is a prime ideal of ZC_3.
(c) Prove that $(7, 2 - \sigma) \neq (7, \sigma + 3)$.

(45) Prove Proposition 2.25.
(46) Let R be a commutative ring with unity. By Proposition 2.23, every proper ideal of R is contained in a maximal ideal. This is proved using Zorn's lemma. Show that the same result holds *without using Zorn's lemma* if R is a Noetherian commutative ring with unity.
(47) Prove Proposition 2.27.
(48) Let R be a ring, let G be a finite group and let RG be the group ring. Let $S = \{r \sum_{g \in G} g : r \in R\}$. Prove that S is an ideal of RG.
(49) Let R be a ring and suppose that

$$I_1 \subseteq I_2 \subseteq \cdots \subseteq I_j \subseteq I_{j+1} \subseteq \cdots$$

is an ascending sequence of ideals. Show that $\bigcup_{j=1}^{\infty} I_j$ is an ideal of R.
(50) Let M_1 and M_2 be maximal ideals of R. Show that either $M_1 + M_2 = R$ or $M_1 = M_2$.
(51) Prove that the ideal $(x, x^2 + 2)$ in $Z[x]$ is not principal.

(52) Let K be a field extension of $\mathbb{Q}$. Let $p(x)$ be an irreducible polynomial in $K[x]$. Prove that $p(x)$ has distinct roots.

(53) Let $p(x) = a_0 + a_1x + a_2x^2$ be a quadratic polynomial in $\mathbb{Q}[x]$. Suppose that $\gcd(p(x), x^3 - 1) = 1$. Show that $p(\sigma)$ is a unit in the group ring $\mathbb{Q}C_3$, $C_3 = \langle \sigma \rangle$.

Exercises for §2.5

(54) Find a group that is isomorphic to the additive group of the quotient ring $Z_{10}/\{0, 5\}$.

(55) Let N be the principal ideal of $Z \times Z$ generated by $(2, 4)$. List the elements in the quotient ring $(Z \times Z)/N$.

(56) In the ring $\mathbb{Q}[x]$, let $N = (x^2 - 3x + 2)$ be the principal ideal generated by $x^2 - 3x + 2$. Prove that $((x - 1) + N)^2 = (x - 1) + N$ in $\mathbb{Q}[x]/N$.

(57) Prove that $Z_5[x]/(x^3 + 2x + 1)$ is a field.

(58) Prove that $\mathbb{Q}[x]/(x^3 - x^2 + 2x - 1)$ is an integral domain.

(59) Let C_3 denote the cyclic group of order 3 generated by σ. Prove that $ZC_3/(7, 2 - \sigma)$ is a field.

(60) Let R be a ring and let I be an ideal of R which is not maximal. Show that R/I has at least three ideals.

(61) Let I be an ideal of R with quotient ring R/I. Let a be a unit of R. Show that $a + I$ is a unit of R/I.

(62) Suppose $a + I$ is a unit in the quotient ring R/I. Give an example which shows that a need not be a unit of R.

(63) Let K be a field and let $K[w, x, y, z]$ be the ring of polynomials in the indeterminates w, x, y, z. Prove that the quotient ring $K[w, x, y, z]/(wx - yz)$ is not a UFD. Is it an integral domain?

(64) True or False: $\mathbb{R} \cong \mathbb{Q}$ as rings.

(65) Consider the rings $\mathbb{Q}$ and $\mathbb{Q} \times \mathbb{Q}$.

  (a) Show that there exists a bijective map $f : \mathbb{Q} \to \mathbb{Q} \times \mathbb{Q}$.
  (b) True or False: $\mathbb{Q} \cong \mathbb{Q} \times \mathbb{Q}$ as rings.

(66) Let $\phi : R \to R'$ be a surjective homomorphism of commutative rings. If both rings R, R' have unity elements $1_R, 1_{R'}$, prove that $\phi(1_R) = 1_{R'}$.

(67) Let R and R' be commutative rings with unity. Find an example of a ring homomorphism $\phi : R \to R'$ for which $\phi(1_R) \neq 1_{R'}$.

(68) Let $\phi : \mathbb{Q} \to \mathbb{C}$ be an injective homomorphism of commutative rings with unity. Show that $\phi(x) = x, \forall x \in \mathbb{Q}$. (Hint: First show that $\phi|_Z : Z \to \mathbb{C}$ maps n to n, for $n \in Z$.)

(69) Let $\phi : F \to R$ be a homomorphism of rings with F an infinite field and R finite. Show that $\phi(x) = 0$ for all $x \in F$.
(70) Let $\phi : \mathbb{R} \to \mathbb{C}$ be the function defined by $\phi(r) = r\mathrm{i}$, $\mathrm{i} = \sqrt{-1}$, for all $r \in \mathbb{R}$. Determine whether ϕ is a ring homomorphism.
(71) Compute the kernel of the evaluation homomorphism $\phi_2 : Z_{29}[x] \to Z_{29}$
(72) Compute the kernel of the evaluation homomorphism $\phi_{1+\sqrt{2}} : \mathbb{Q}[x] \to \mathbb{R}$.
(73) Let R be a commutative ring with unity and let G be a finite group. Let $\epsilon : RG \to R$ be the map defined as $\epsilon(\sum_{g \in G} a_g g) = \sum_{g \in G} a_g$.

(a) Show that ϵ is a surjective ring homomorphism.
(b) Find $\ker(\epsilon)$.

(74) Let H be a normal subgroup of G. Let $N = G/H$ be the quotient group, and let $\phi : RG \to RN$ be the map of group rings defined by $g \mapsto gH$. Show that ϕ is a surjective ring homomorphism.
(75) Compute the characteristic of the ring $Z_3[x]$.
(76) Suppose that R is an integral domain with $r = \text{char}(R) > 0$. Prove that r is a prime number.
(77) Let p be a prime number, let $n \geq 1$ be an integer and let F be a field with $\text{char}(F) = p$. Prove that the map $\phi : F \to F$ defined as $\phi(x) = x^{p^n}, \forall x \in F$, is a homomorphism of fields.
(78) Give another proof of the Eisenstein Criterion (Proposition 2.8) by completing the steps below. As in the proof already given, we assume that the degree n polynomial $f(x)$ factors as $f(x) = g(x)h(x)$ in $Z[x]$, with $\deg(g(x) = k$, $\deg(h(x)) = l$, $k, l \geq 1$.

(a) Let $\phi : Z[x] \to Z_p[x]$ be the map defined as $\phi(p(x)) = \overline{p}(x)$, where the coefficients of $\overline{p}(x)$ are those of $p(x)$ taken modulo p. Show that ϕ is a homomorphism of rings.
(b) Apply (a) to show that $\overline{g}(x) = x^k$ and $\overline{h}(x) = x^l$.
(c) From (b) conclude that $f(0) \equiv 0 \bmod p^2$, a contradiction.

Exercises for §2.6

(79) Let R be a commutative ring with unity. Show that the set of nilpotents elements of R is an ideal of R.
(80) Find all of the nilpotent elements in the group ring Z_3C_3.
(81) List all of the elements in the localization $(Z_6)_2$.

(82) Show that x is a non-nilpotent element of $Z[x]$. List four elements in the localization $Z[x]_x$.

(83) Prove that (x) is a prime ideal of $Z[x]$. Prove that the localization $Z[x]_{(x)}$ is not the same as $Z[x]_x$.

(84) Let R be a Noetherian commutative ring with unity and let S be a multiplicative subset of R. Show that the localization $S^{-1}R$ is Noetherian.

(85) Let R be a local ring with maximal ideal m. Let $b \in R$ with $b \notin m$. Prove that b is a unit of R.

(86) Let $S = \{1, 3, 9, 7\} \subseteq Z_{10}$. Compute the kernel of $\lambda : Z_{10} \to S^{-1}Z_{10}$, $a \mapsto a/1$.

(87) Prove that every non-zero integer is a unit in $Z_{(p)}$ for an infinite number of primes p.

(88) Verify the formulas:

(a) $Z = \bigcap_{p \text{ prime}} Z_{(p)}$,

(b) $Z_f = \bigcap_{p \nmid f} Z_{(p)}$ for f a non-nilpotent integer.

Exercises for §2.7

(89) Show that $\sqrt{2}$ is the limit of a $|\ |$-Cauchy sequence in $\mathbb{Q}$.

(90) Prove that $|\ |_p$ is an absolute value on $\mathbb{Q}$.

(91) Prove that the sequence $a_n = \frac{1}{n}$ is a $|\ |$-Cauchy sequence in $\mathbb{Q}$. Is $\{a_n\}$ $|\ |_3$-Cauchy?

(92) Prove Proposition 2.48.

(93) Suppose $\mathbb{Q}$ is endowed with the p-adic absolute value. Let x, y, z be distinct non-zero elements in $\mathbb{Q}$. Let $a = |x - z|_p$, $b = |x - y|_p$, $c = |y - z|_p$.

(a) Show that there exists a triangle whose sides have lengths a, b, c.

(b) Show that any triangle constructed in (a) is isosceles.

(94) Construct a root of $x^2 - 2$ in $\mathbb{Q}_7$.

(95) Referring to Proposition 2.51, show that $\mathcal{N}$ is an ideal of $\mathcal{C}$.

(96) Prove that $Z_{(p)} = \mathbb{Q} \cap \mathbb{Z}_p$.

(97) Let x be an element of $\mathbb{Z}_p$ with $x \in Z_{(p)}$. Prove that there exist $a, b \in Z$ such that the equation $x = \sum_{i=0}^{\infty} ar^i$ holds in $\mathbb{Z}_p$ where $r = bp$.

(98) Prove that $\mathbb{Z}_p$ is a PID.

(99) Verify that

$$0 = 1 + 1 + 2 + 4 + 8 + 16 + \cdots$$

in $\mathbb{Z}_2$.

(100) Let $\{p_n\}$ denote the sequence of prime numbers $2, 3, 5, 7, 11, 13, \ldots$. Let m be a positive integer.

(a) Compute $\lim_{n\to\infty} |m|_{p_n}$.

(b) Discuss the value of $\lim_{m\to\infty} \lim_{n\to\infty} |m|_{p_n}$.

(101) Show that for each $x \in \mathbb{Q}$, $|x| \cdot \prod_{p \text{ prime}} |x|_p = 1$.

Questions for Further Study

(1) Let R be a commutative ring with unity and let G, G' be finite groups of order n.

(a) If $G \cong G'$, show that $RG \cong RG'$ as rings.

(b) Prove that the converse of (a) is false, that is, find two groups G, G' for which $RG \cong RG'$, yet $G \ncong G'$.

Chapter 3

Modules

In this chapter we introduce the concept of a module over a commutative ring with unity. We begin with the familiar notion of a vector space, which is a module over a field. We include both finite and infinite dimensional vector spaces, subspaces and quotient spaces, linear transformations of vector spaces and the linear dual of a vector space. We generalize the definition of a vector space V over a field K to that of a module M over a commutative ring with unity R. An underlying theme is to determine which theorems and properties of vector spaces carry over to the more general setting of modules. For instance, the concept of a finite dimensional vector space over K generalizes to a free R-module M of finite rank. The notion of a free R-module is then generalized to that of a projective module; we show that a finitely generated and projective module over a local ring is free of finite rank. Next, we define tensor products of two modules, specialize to tensor products of vector spaces, and relate these to the linear dual.

We next consider algebras A over a ring R, an object that is a ring A as well as an R-module, where the scalar multiplication is given through a ring homomorphism $\lambda : R \to A$. We specialize to commutative R-algebras, and show that every finitely generated R-algebra is the quotient of a polynomial ring. We compute all of the R-algebra homomorphisms $A \to B$ where A is finitely generated.

Finally, we discuss bilinear forms on a finite dimensional vector space V over the field K, where K is the field of fractions of an integral domain R. For an R-submodule M of V, which is free over R and which satisfies $KM = V$, we define the dual module of M^D and the discriminants $\mathrm{disc}(M)$ and $\mathrm{disc}(M^D)$. As an example we compute $\mathrm{disc}(RG)$ and $\mathrm{disc}(RG^D)$, where G is the Klein 4-group. We shall use discriminants in Chapter 4.

3.1 Vector Spaces

In this section we begin with the definition of a vector space V over a field K. Vector spaces (at least finite dimensional vector spaces) should be familiar to readers of this book. We review linear independence of vectors, spanning sets, bases, and prove that every vector space V admits a basis. We define finite and infinite dimensional vector spaces, and consider subspaces and quotient spaces. Next, we turn to linear transformations of vector spaces, specializing to the linear dual.

* * *

Definition 3.1. Let K be a field. A **vector space over** K is an additive abelian group V together with a scalar multiplication $K \times V \to V$, denoted by $(r, v) \mapsto rv$, that satisfies, for all $r, s \in K$, $v, w \in V$,

(i) $r(v + w) = rv + rw$,

(ii) $(r + s)v = rv + sv$,

(iii) $(rs)v = r(sv)$,

(iv) $1v = v$.

An element $v \in V$ is a **vector**. Note that the conditions of the definition imply that $0v = 0$, $\forall v$.

The most familiar example of a vector space is Euclidean n-space $\mathbb{R}^n$ which consists of n-tuples $(a_1, a_2, \ldots, a_n)$, $a_i \in \mathbb{R}$, equipped with the usual vector addition and scalar multiplication. Other examples include the K-vector space of polynomials $K[x]$ with scalar multiplication $K \times K[x] \to K[x]$ defined as $(r, f(x)) \mapsto rf(x)$, for $r \in K$, $f(x) \in K[x]$, and the group ring KG, with scalar multiplication $K \times KG \to KG$ defined by $(r, g) \mapsto rg$ for $r \in K$, $g \in G$. In the field extension L/K, L is a vector space over K with scalar multiplication $K \times L \to L$ defined as $(x, y) \mapsto xy$ for $x \in K$, $y \in L$.

Let V be a vector space over a field K and let $S = \{v_\alpha\}_{\alpha \in I}$ be a subset of V indexed by the set I. Then S is **linearly independent** if

$$\sum_{\alpha \in I} r_\alpha v_\alpha = 0, \quad r_\alpha \in K,$$

where $r_\alpha = 0$ for all but a finite number of α, implies that $r_\alpha = 0$ for all $\alpha \in I$. If $v \neq 0$, then the singleton subset $\{v\}$ is a linearly independent subset of V.

Definition 3.2. Let $S = \{v_\alpha\}_{\alpha \in I}$ be a subset of a K-vector space V. Then S is a **generating set** for V (or S **spans** V) if each $v \in V$ has a representation $v = \sum_{\alpha \in I} r_\alpha v_\alpha$ for $r_\alpha \in K$ with $r_\alpha = 0$ for all but a finite number of indices α.

A vector space is **finitely generated** if there exists a generating set S that is finite. For example, $\mathbb{R}^n$ is a finitely generated vector space over $\mathbb{R}$.

Definition 3.3. Let $S = \{v_\alpha\}_{\alpha \in I}$ be a subset of the K-vector space V. Then S is a **basis** for V if each $v \in V$ has a unique representation $v = \sum_{\alpha \in I} r_\alpha v_\alpha$ for $r_\alpha \in K$ with $r_\alpha = 0$ for all but a finite number of indices α.

Proposition 3.1. *Let V be a K-vector space. The subset $S = \{v_\alpha\}_{\alpha \in I}$ is a basis for V if and only if S is a linearly independent spanning set for V.*

Proof. Assume that $S = \{v_\alpha\}_{\alpha \in I}$ is a basis. Then certainly S spans V. Now suppose that $\sum_{\alpha \in I} r_\alpha v_\alpha = 0$, for $r_\alpha \in K$. Then $\sum_{\alpha \in I} r_\alpha v_\alpha$ is a representation of the zero vector, but since $\sum_{\alpha \in I} 0 v_\alpha$ also represents zero, we conclude that $r_\alpha = 0, \forall \alpha$. Thus S is linearly independent.

For the converse suppose that $v = \sum_{\alpha \in I} r_\alpha v_\alpha = \sum_{\alpha \in I} s_\alpha v_\alpha$ for some $v \in V$. Then $\sum_{\alpha \in I} (r_\alpha - s_\alpha) v_\alpha = 0$, hence $r_\alpha - s_\alpha = 0, \forall \alpha \in I$, and so, S is a basis for V. □

Proposition 3.2. *Let V be a vector space over a field K. Then there exists a basis for V.*

Proof. We employ a Zorn's Lemma argument (Chapter 2, §2.4). Let S be a generating set for V over K (such S always exists, one could take $S = V$). Assume that $S \neq \{0\}$, so that S contains a linearly independent subset $\{v\}$ of V. Now let $\mathcal{T}$ denote the collection of linearly independent subsets of S that contain $\{v\}$. Then $\mathcal{T}$ is a non-empty set partially ordered under set inclusion. Moreover, every chain $\{T_i\}$ in $\mathcal{T}$ has an upper bound $T = \bigcup_i T_i$. Thus by Zorn's Lemma, $\mathcal{T}$ has a maximal element B. By its construction, B is a linearly independent subset of V.

We claim that B is a basis for V. To this end let W be the subset of V consisting of all vectors of the form $\sum_{b \in B} r_b b$, where $r_b = 0$ for all but a finite number of $b \in B$ (W is the subspace of V generated by B). If $V = W$, then B is a basis for V. Else, there exists an element $x \in V \backslash W$. Suppose that $\sum_{b \in B} r_b b + rx = 0$ for some $r_b, r \in K$. If $r \neq 0$, then $x = -\sum_{b \in B} (r_b / r) b \in W$, which is a contradiction. Consequently,

$\sum_{b\in B} r_b b + rx = 0$ implies that $r_b = r = 0, \forall b \in B$, and so, $B \cup \{x\}$ is a linearly independent subset of V, containing $\{v\}$ and containing B as a proper subset. This contradicts the maximality of B. □

Moreover, if the vector space V is finitely generated by a set with n elements, then there is a basis for V that contains no more than n elements.

Proposition 3.3. *Let V be a vector space over K that is finitely generated by a set S with n elements. Then there exists a basis for V that contains $m \le n$ elements.*

Proof. Let $S = \{v_1, v_2, \ldots, v_n\}$ be a generating set for V over K, and let B be the smallest subset of S for which B generates V over K. Renumbering the elements of S if necessary, write $B = \{v_1, v_2, \ldots, v_m\}$ with $m \le n$. We claim that B is a basis for V. By way of contradiction, suppose there exists an element $v \in V$ for which

$$v = a_1v_1 + a_2v_2 + \cdots + a_mv_m, \quad a_i \in K,$$
$$v = b_1v_1 + b_2v_2 + \cdots + b_mv_m, \quad b_i \in K,$$

are two distinct representations of v. Then there exists an integer i' for which $a_{i'} \neq b_{i'}$. Thus

$$(a_{i'} - b_{i'})v_{i'} = \sum_{\substack{i=1,\\ i\neq i'}}^{m} (b_i - a_i)v_i,$$

and so,

$$v_{i'} = \sum_{\substack{i=1,\\ i\neq i'}}^{m} \left(\frac{b_i - a_i}{a_{i'} - b_{i'}}\right) v_i.$$

Hence $B\backslash\{v_{i'}\}$ generates V, which contradicts the minimality of B. □

Thus every finitely generated vector space admits a finite basis. The next proposition shows that the number of elements in any finite basis is the same.

Proposition 3.4. *Let V be a finitely generated vector space. Then any two bases for V have the same number of elements.*

Proof. Suppose that $\{b_1, b_2, \ldots, b_m\}$ and $\{c_1, c_2, \ldots, c_n\}$ are two different bases for V with $m \neq n$. Without loss of generality, we may assume that $m < n$. For $1 \leq j \leq n$, write $c_j = \sum_{i=1}^{m} a_{i,j} b_i$ for unique elements $a_{i,j} \in K$. Let $A = (a_{i,j})$, and let $A' = (a'_{i,j})$ denote the reduced row echelon form of A. Since $m < n$, there exists an integer j^*, $m < j^* \leq n$ so that

$$c_{j^*} = a'_{1,j^*} c_1 + a'_{2,j^*} c_2 + \cdots + a'_{m,j^*} c_m.$$

Hence $\{c_1, c_2, \ldots, c_n\}$ is not linearly independent, a contradiction. □

In view of Proposition 3.4, if V finitely generated, then we define the **dimension of** V to be the number of elements in any basis for V over K. We denote the dimension of V by $\dim(V)$; we say that the vector space V is **finite dimensional**.

If L/K is a field extension in which L is finitely generated over K, hence $\dim(L) = n$ for some n, then L/K is a **finite extension of fields**. If L/K is a finite extension of fields, then the **degree** of L over K, denoted as $[L : K]$ is the dimension of L as a vector space over K.

Note that the **zero vector space** $\{0\}$ is finite dimensional with $\dim(\{0\}) = 0$. If the vector space V is not finitely generated and hence does not admit a finite basis, then V is **infinite dimensional**.

Let V be a vector space over K. A **subspace** W **of** V is an additive subgroup of V for which $rw \in W$ for all $r \in K$, $w \in W$. The subspace W is a K-vector space whose scalar multiplication is the scalar multiplication of V restricted to W. One can always construct a subspace of V as follows. Let $S = \{v_\alpha\}_{\alpha \in I}$ be a subset of V and let W consist of all elements of V the form $\sum_{\alpha \in I} r_\alpha v_\alpha$ where $r_\alpha \in K$ and $r_\alpha = 0$ for all but a finite number of α. Then W is a subspace of V called the **subspace of** V **generated (or spanned) by** S.

Proposition 3.5. *Suppose V is a finite dimensional vector space and W is a subspace of V. Then $\dim(W) \leq \dim(V)$.*

Proof. As a vector space, W is finitely generated by a set containing $\leq \dim(V)$ elements. By Proposition 3.3, there exists a basis for W with $\leq \dim(V)$ elements. □

Let W be a subspace of V. Since W is a normal subgroup of V, we may form the quotient group V/W. Now since $r(v + W) \subseteq rv + W$ for all $r \in K$, $v \in V$, the quotient group V/W is a vector space with scalar multiplication $K \times V/W \to V/W$ defined as $r(v + W) = rv + W$. The vector space V/W is the **quotient space of** V **by** W. (One should check that the map $K \times V/W \to V/W$ is well-defined on left cosets.) If V/W is finite dimensional then the subspace W is **cofinite** and $\dim(V/W) < \infty$ is the **codimension of** W.

Definition 3.4. Let V, V' be vector spaces. A **linear transformation** is a map $\phi : V \to V'$ that satisfies, for all $x, y \in V$, $r \in K$,

(i) $\phi(x + y) = \phi(x) + \phi(y)$,

(ii) $\phi(rx) = r\phi(x)$.

The linear transformation $\phi : V \to V'$ is an **isomorphism** of vector spaces if ϕ is a bijection.

Let V be a vector space over a field K. We consider K as a vector space over itself. A **linear functional on** V is a linear transformation $f : V \to K$. The collection of all linear functionals on V is the **linear dual of** V and is denoted as V^*. The linear dual V^* is a vector space over K with vector addition defined as

$$(f + g)(v) = f(v) + g(v),$$

for all $f, g \in V^*$, $v \in V$, and scalar multiplication given as

$$(r, f) \mapsto rf,$$

with $(rf)(v) = rf(v)$ for $r \in K$, $f \in V^*$, $v \in V$.

Let $\phi : V \to W$ be a linear transformation of K-vector spaces. Then ϕ induces a map

$$\phi^* : W^* \to V^*,$$

defined as $\phi^*(f)(v) = f(\phi(v))$ for $f \in W^*$, $v \in V$. The map ϕ^* is the **transpose of** ϕ.

Proposition 3.6. *$\phi^* : W^* \to V^*$ is a linear transformation of vector spaces.*

Proof. Exercise. □

Proposition 3.7. *A finite dimensional vector space and its dual have the same dimension.*

Proof. By hypothesis, V has a finite basis $\{b_1, b_2, \ldots, b_n\}$. We shall construct a K-basis for V^* with n elements. For each i, $1 \leq i \leq n$, define a map $f_i : V \to K$ by

$$f_i(r_1 b_1 + r_2 b_2 + \cdots + r_n b_n) = r_i,$$

for $r_i \in K$, $1 \leq i \leq n$. The map f_i is the i**th coordinate map**. The collection of coordinate maps $\{f_1, f_2, \ldots, f_n\}$ is a subset of V^*. We show that $\{f_i\}$ is a basis for V^*. Let $f \in V^*$. Then

$$f = f(b_1)f_1 + f(b_2)f_2 + \cdots + f(b_n)b_n,$$

and so $\{f_i\}$ generates V^*. Now suppose that

$$r_1 f_1 + r_2 f_2 + \cdots + r_n f_n = 0,$$

for $r_i \in K$, where 0 is the zero functional: $0(v) = 0$ for all $v \in V$. Then for all i,

$$\begin{aligned} 0 &= 0(b_i) \\ &= (r_1 f_1 + r_2 f_2 + \cdots + r_n f_n)(b_i) \\ &= r_i, \end{aligned}$$

and so $\{f_i\}$ is a linearly independent subset of V^*. It follows that $\{f_i\}$ is a K-basis for V^*. □

The collection $\{f_i\}_{i=1}^n$ is the basis for V^* dual to the basis $\{b_i\}_{i=1}^n$ for V; $\{f_i\}$ is the **dual basis** for V^* with respect to $\{b_i\}$.

We next consider a more general construction than the linear dual. Let X be a non-empty set of elements and let $\mathrm{Map}(X, K)$ denote the set of all functions $f : X \to K$ (here, X is not necessarily a vector space and the maps need not be linear transformations). Then $\mathrm{Map}(X, K)$ is a vector space with vector addition defined pointwise: for $f, g \in \mathrm{Map}(X, K)$, $x \in X$, $(f+g)(x) = f(x)+g(x)$ and scalar multiplication given as $(rf)(x) = rf(x)$ for $r \in K$.

Proposition 3.8. *Let W be a subspace of $Map(X, K)$ of finite dimension n. Then there exist elements $x_1, x_2, \ldots, x_n \in X$ and a basis $\{f_1, f_2, \ldots, f_n\}$ of W for which $f_i(x_j) = \delta_{i,j}$, $1 \leq i, j \leq n$, where $\delta_{i,j}$ is Kronecker's delta.*

Proof. We proceed by induction on n. For the trivial case $n = 1$, let $\{h_1\}$ be a basis for W, $\dim(W) = 1$. Since $h_1 \neq 0$, there exists an element $x_1 \in X$ so that $h_1(x_1) \neq 0$. Define a function $f_1 : X \to K$ by the rule $f_1 = h_1(x_1)^{-1}h_1$. Then $\{f_1\}$ is a K-basis for W with $f_1(x_1) = 1$ as required.

For the induction hypothesis, we assume that whenever $W \subseteq \mathrm{Map}(X, K)$ is a subspace of dimension $n - 1$, there exist elements $x_1, x_2, \ldots, x_{n-1} \in X$ and a basis $\{f_1, f_2, \ldots, f_{n-1}\}$ for W so that $f_i(x_j) = \delta_{i,j}$, $1 \leq i, j \leq n - 1$.

Now suppose that $\dim(W) = n$ and let $\{h_1, h_2, \ldots, h_n\}$ be a basis for W. Let W' be the subspace of $\mathrm{Map}(X, K)$ spanned by $\{h_1, h_2, \ldots, h_{n-1}\}$. Clearly, $\dim(W') = n - 1$ and so by the induction hypothesis, there exist elements $x_1, x_2, \ldots, x_{n-1} \in X$ and a basis $\{f_1', f_2', \ldots, f_{n-1}'\}$ for W' so that $f_i'(x_j) = \delta_{i,j}$, $1 \le i, j \le n - 1$. Let

$$g = h_n - h_n(x_1)f_1' - h_n(x_2)f_2' - \cdots - h_n(x_{n-1})f_{n-1}'.$$

Suppose that $g = 0$. Then

$$h_n(x) = h_n(x_1)f_1'(x) + h_n(x_2)f_2'(x) + \cdots + h_n(x_{n-1})f_{n-1}'(x),$$

for all $x \in X$, and so, $h_n \in W'$. It follows that $\dim(W) = n - 1$, a contradiction. Hence, there exists an element $x_n \in X$ with $g(x_n) \neq 0$. Let $f_n = g(x_n)^{-1}g$ and $f_i = f_i' - f_i'(x_n)f_n$ for $1 \le i \le n-1$. Then $f_i(x_j) = \delta_{i,j}$, $1 \le i, j \le n$. We claim that $\{f_i\}_{i=1}^n$ is a basis for W. To this end, suppose there exists $r_1, r_2, \ldots, r_n \in K$ for which

$$r_1 f_1 + r_2 f_2 + \cdots + r_n f_n = 0.$$

Then evaluation at x_i yields $r_i = 0$, and hence, $\{f_1, f_2, \ldots, f_n\}$ is linearly independent. Note that $h_i \in \mathrm{span}(f_1, f_2, \ldots, f_n)$ for $1 \le i \le n - 1$ since $f_i' = f_i + f_i'(x_n)f_n$ for $1 \le i \le n - 1$. Moreover, $h_n \in \mathrm{span}(f_1, f_2, \ldots, f_n)$ since

$$\begin{aligned}
h_n &= g + h_n(x_1)f_1' + h_n(x_2)f_2' + \cdots + h_n(x_{n-1})f_{n-1}' \\
&= g(x_n)f_n + h_n(x_1)f_1' + h_n(x_2)f_2' + \cdots + h_n(x_{n-1})f_{n-1}' \\
&= g(x_n)f_n + h_n(x_1)(f_1 + f_1'(x_n)f_n) + h_n(x_2)(f_2 + f_2'(x_n)f_n) \\
&\quad + \cdots + h_n(x_{n-1})(f_{n-1} + f_{n-1}'(x_n)f_n).
\end{aligned}$$

Consequently, $W = \mathrm{span}(f_1, f_2, \ldots, f_n)$, which says that $\{f_1, f_2, \ldots, f_n\}$ is a basis for W, as required. □

Of course, in the case that $X = V$, $W = V^*$ is a subspace of $\mathrm{Map}(X, K)$. Furthermore, if $\dim(V) = n$, then $\dim(V^*) = n$ and Proposition 3.8 applies to yield elements $x_1, x_2, \ldots, x_n \in V$ and a basis $\{f_1, f_2, \ldots, f_n\}$ for V^* with $f_i(x_j) = \delta_{i,j}$. Proposition 3.7, however, asserts that $\{x_j\}$ is a basis for V; $\{f_i\}$ is the dual basis.

3.2 Modules

In this section we generalize the definition of a vector space V over a field K to that of a module M over a commutative ring with unity R. An underlying theme here is to determine which properties of vector spaces carry over to the more general setting of modules. We consider generating sets for R-modules, bases for free modules, as well as submodules and quotient modules. As with vector spaces (Proposition 3.4) the rank of a free R-module is well-defined. Next, we study homomorphisms of modules and give R-module analogs for the First Isomorphism Theorem (for groups) and the Universal Mapping Property for Kernels. We show that the subspace property of vector spaces (Proposition 3.5) generalizes to PIDs and a still weaker version of Proposition 3.5 is valid for Noetherian rings.

* * *

Throughout this section R is a commutative ring with unity.

Definition 3.5. Let R be a commutative ring with unity. Then a **left module over** R is an additive abelian group M together with a scalar multiplication $R \times M \to M$, denoted by $(r, m) \mapsto rm$, that satisfies, for all $r, s \in R$, $m, n \in M$,

(i) $r(m+n) = rm + rn$,

(ii) $(r+s)m = rm + sm$,

(iii) $(rs)m = r(sm)$,

(iv) $1m = m$.

Note that the conditions of the definition imply that $0m = 0$, $\forall m$.

Of course, a vector space V over a field K is a K-module. Any ideal I of a ring R is an R-module, the scalar multiplication being the multiplication in R – specifically R is a module over itself. Any additive abelian group G is a Z-module with scalar multiplication defined as $0g = 0$, $ng = \underbrace{g + g + \cdots + g}_{n}$, for $n > 0$, and $ng = \underbrace{(-g) + (-g) + \cdots + (-g)}_{|n|}$, for $n < 0$. The ring of polynomials $R[x]$ is an R-module with scalar multiplication $R \times R[x] \to R[x]$ defined as $(r, f(x)) \mapsto rf(x)$, for $r \in R$, $f(x) \in R[x]$. The group ring RG is an R-module; the scalar multiplication $R \times RG \to RG$ is given by $(r, g) = rg$.

Let M and N be R-modules. The **direct sum** $M \oplus N$ is the R-module consisting of all pairs $(a, b) \in M \times N$ where addition is defined as $(a, b) + (c, d) = (a + c, b + d)$ and scalar multiplication is given as $r(a, b) = (ra, rb)$ for $r \in R$. More generally, let $M_1, M_2, \ldots, M_k$ be a finite collection of R-modules. The direct sum $M_1 \oplus M_2 \oplus \cdots \oplus M_k$ is the R-module consisting of all k-tuples in $\prod_{i=1}^{k} M_i$ where addition is given as

$$(a_1, a_2, \ldots, a_k) + (b_1, b_2, \ldots, b_k) = (a_1 + b_1, a_2 + b_2, \ldots, a_k + b_k),$$

and scalar multiplication is given as $r(a_1, a_2, \ldots, a_k) = (ra_1, ra_2, \ldots, ra_k)$.

Definition 3.6. Let $S = \{m_\alpha\}_{\alpha \in I}$ be a subset of the R-module M. Then S is a **generating set** for M if each $m \in M$ has a representation $m = \sum_{\alpha \in I} r_\alpha m_\alpha$ for $r_\alpha \in R$ with $r_\alpha = 0$ for all but a finite number of indices α.

Definition 3.7. The R-module M is **free** if there exists a subset $B = \{m_\alpha\}_{\alpha \in I}$ of M for which each $m \in M$ has a unique representation $m = \sum_{\alpha \in I} r_\alpha m_\alpha$ for $r_\alpha \in R$ with $r_\alpha = 0$ for all but a finite number of indices α. The subset B is a **basis for M over R**.

For example, $R[x]$ is free over R on the basis $B = \{1, x, x^2, x^3, \ldots\}$. Here is a general construction of a free module. Let S be a non-empty set. The **free R module on S**, $F\langle S \rangle$ is the collection of all formal sums

$$\sum_{s \in S} r_s s$$

where $r_s = 0$ for all but a finite number of elements $s \in S$. The addition in $F\langle S \rangle$ is defined as

$$\sum_{s \in S} r_s s + \sum_{s \in S} t_s s = \sum_{s \in S} (r_s + t_s) s,$$

and scalar mulitplication is given as

$$t \cdot \sum_{s \in S} r_s s = \sum_{s \in S} (tr_s) s,$$

for $t \in R$. Clearly, S is a basis for $F\langle S \rangle$.

Definition 3.8. The R-module M is **finitely generated** if there exists a finite subset $S = \{m_1, m_2, \ldots, m_k\}$ of M for which each element $m \in M$ has a representation $m = \sum_{i=1}^{k} r_i m_i$ for $r_i \in R$.

Definition 3.9. M is a **free R-module of rank k** if there exists a finite subset $B = \{m_1, m_2, \ldots, m_k\}$ of M for which each element $m \in M$ has a unique representation $m = \sum_{i=1}^{k} r_i m_i$ for $r_i \in R$; B is a **basis** for M. If $M = \{0\}$, then M has rank 0.

For example, the ring R is a free rank one R-module on the basis $\{1_R\}$. Also, the group ring RG for G finite, is a free R-module of rank $|G|$, an R-basis being the elements of G. Likewise, the collection $\mathrm{Mat}_2(R)$ of 2×2 matrices with entries in R is free over R of rank 4 on the basis

$$\left\{ \begin{pmatrix} 1 & 0 \\ 0 & 0 \end{pmatrix}, \begin{pmatrix} 0 & 1 \\ 0 & 0 \end{pmatrix}, \begin{pmatrix} 0 & 0 \\ 1 & 0 \end{pmatrix}, \begin{pmatrix} 0 & 0 \\ 0 & 1 \end{pmatrix} \right\}.$$

The notion of finite rank generalizes the definition of dimension to modules: a vector space V with $\dim(V) = n$ is a free K-module of rank n.

In contrast to vector spaces (cf. Proposition 3.3), however, a finitely generated R-module M need not be free of finite rank. For instance, the additive group Z_3, is finitely generated as a Z-module, but not free over Z: indeed $m = 1m = 4m = 7m$ for any $m \in Z_3$.

Let M be an R-module. An **R-submodule N of M** is an additive subgroup N of M for which $rn \in N$ for $r \in R$, $n \in N$. Note that a K-submodule W of the vector space V is a subspace of V. One can always construct a submodule of M as follows. Let $S = \{m_\alpha\}_{\alpha \in I}$ be a subset of M and let N consist of all elements of M the form $\sum_{\alpha \in I} r_\alpha m_\alpha$ where $r_\alpha \in R$ and $r_\alpha = 0$ for all but a finite number of α. Then N is an R-submodule of M called the **submodule of M generated by S**.

Let N be a submodule of M. Since N is a normal subgroup of M, we may form the quotient group M/N. Now since $r(a + N) \subseteq ra + N$ for all $r \in R$, the quotient group M/N is an R-module with scalar multiplication $R \times R/N \to R/N$ defined as $r(a + N) = ra + N$. The R-module M/N is the **quotient module of M by N**.

As with vector spaces (cf. Proposition 3.4) the rank of a free R-module is well-defined.

Proposition 3.9. *Let M be a free R-module of finite rank. Then any two bases for M have the same number of elements.*

Proof. By definition, M has a finite basis $B = \{m_1, m_2, \ldots, m_k\}$ for some integer k. Let $C = \{c_1, c_2, \ldots, c_n\}$ be any other basis for M. By Proposition 2.23, R has a maximal ideal J. Now

$$JM = \{\sum rm : \ r \in J, m \in M\}$$

is an R-submodule of M. The quotient module M/JM is a finitely generated vector space over the field R/J. Note that $\overline{B} = \{m_i + JM : \ m_i \in B\}$ and $\overline{C} = \{c_i + JM : \ c_i \in C\}$ are bases for M/JM over R/J. So by Proposition 3.4, $|\overline{B}| = |\overline{C}|$. It follows that $k = |B| = |C|$. $\square$

We next describe the basic map between two R-modules.

Definition 3.10. An R-module homomorphism is a map $\phi : M \to M'$ of R-modules that satisfies, for all $x, y \in M$, $r \in R$,

(i) $\phi(x + y) = \phi(x) + \phi(y)$,

(ii) $\phi(rx) = r\phi(x)$.

The R-module homomorphism $\phi : M \to M'$ is an **isomorphism** of R-modules if ϕ is a bijection. In this case we write $M \cong M'$ as R-modules. If M is a free R-module of rank k, then $M \cong \underbrace{R \oplus R \oplus \cdots \oplus R}_{k}$. To see this, let $x_1, x_2, \ldots, x_k$ be a basis for M and note that the map $x_i \mapsto (0, \ldots, 0, 1, 0, \ldots, 0)$, 1 in the ith place, 0's elsewhere, is an R-module isomorphism.

Condition (i) of Definition 3.10 says that an R-module homomorphism is a homomorphism of additive abelian groups. The **kernel** of the R-module homomorphism $\phi : M \to M'$ is its kernel as a homomorphism of additive abelian groups, that is, $\ker(\phi) = \{x \in M : \phi(x) = 0\}$. Let $N = \ker(\phi)$. Since $\phi(rx) = r\phi(x) = r0 = 0$ for $r \in R$, $x \in N$, the kernel is an R-submodule of M.

We have easy analogs for Proposition 1.21 and Proposition 1.22.

Proposition 3.10. *Let N be a submodule of M. Then the map $\phi : M \to M/N$ defined by $a \mapsto a + N$ is a surjective homomorphism of R-modules with kernel N.*

Proof. By Proposition 1.21, ϕ is a homomorphism of additive groups, so we only need to check that ϕ respects scalar multiplication, but this is easy: for $r \in R$, $x \in M$, $\phi(rx) = rx + N = r(x + N) = r\phi(x)$. □

Proposition 3.11. *Let $\phi : M \to M'$ be a homomorphism of R-modules with kernel N. Then the map $\gamma : R/N \to \phi(M)$ defined by $\gamma(a+N) = \phi(a)$ is an isomorphism of R-modules.*

Proof. By Proposition 1.22, ϕ is an isomorphism of additive groups, so we only need to check that ϕ respects scalar multiplication, but this is easy: for $r \in R$, $x \in M$, $\gamma(r(x+N)) = \gamma(rx+N) = \phi(rx) = r\phi(x) = r\gamma(x+N)$. □

Here is an application of Proposition 3.11.

Proposition 3.12. *Every finitely generated R-module M is the quotient of a free module of finite rank.*

Proof. Let $\{m_i\}_{i=1}^k$ denote a set of generators for M, let $x_1, x_2, \ldots, x_k$ be indeterminates and let F denote the collection of all sums of the form $\sum_{i=1}^k r_i x_i$ with $r_i \in R$. Then F is a free R-module with basis $\{x_i\}$. There exists a surjective homomorphism of R-modules $\phi : F \to M$ defined as

$$\phi(r_1x_1 + r_2x_2 + \cdots + r_kx_k) = r_1m_1 + r_2m_2 + \cdots + r_kx_k$$

for $r_1, r_2, \ldots, r_k \in R$. Let $N = \ker(\phi)$. Now F/N is a quotient module isomorphic to $\phi(F) = M$, since ϕ is a surjection. □

To illustrate Proposition 3.12, let $R = Z[(1+\sqrt{-23})/2]$. (As we shall see in Chapter 4, R is the ring of integers of the finite field extension $\mathbb{Q}(\sqrt{-23})/\mathbb{Q}$.) Let

$$M = 3R + (2+\sqrt{-23})R$$

be the ideal of R generated by 3 and $2+\sqrt{-23}$; M is a finitely generated module over R. Let F denote the free R-module on the basis $\{x_1, x_2\}$, F is a free rank 2 R-module. Now, there exists a surjective homomorphism of R-modules $\phi : F \to M$ defined as $x_1 \mapsto 3$, $x_2 \mapsto 2+\sqrt{-23}$. Since ϕ is surjective, $F/N \cong M$, as required. Note that $N = \ker(\phi)$ is non-zero: it contains the element $-9x_1 + (2-\sqrt{-23})x_2$. Moreover, R is not a PID and M is not a principal ideal. Consequently, M is not a free rank one R-module.

We have a "universal mapping property for kernels" as in group and ring theory.

Proposition 3.13 (Universal Mapping Property for Kernels). *Let $\phi : M \to M'$ be a homomorphism of R-modules with $N = ker(\phi)$. Suppose that K is a submodule of M with $K \subseteq N$. Then there exists a surjective homomorphism of R-modules $\psi : M/K \to \phi(M)$ defined by $\psi(a+K) = \phi(a)$.*

Proof. Since we already know that ψ is a surjective homomorphism of additive abelian groups, we only need to check that ψ respects scalar multiplication. Let $a + K \in M/K$, $r \in R$. Then

$$\psi(r(a+K)) = \psi(ra+K) = \phi(ra) = r\phi(a) = r\psi(a+K).$$

□

Let $\phi : M \to M'$ be a homomorphism of R-modules with $N = \ker(\phi)$. The Universal Mapping Property for Kernels (UMPK) says that given a submodule K of M with $K \subseteq N$, there exists an R-module homomorphism $\psi : M/K \to M'$ so that

$$\psi s = \phi$$

where $s : M \to M/K$ is the canonical surjection; we say that ϕ "factors through" M/K and the following diagram commutes:

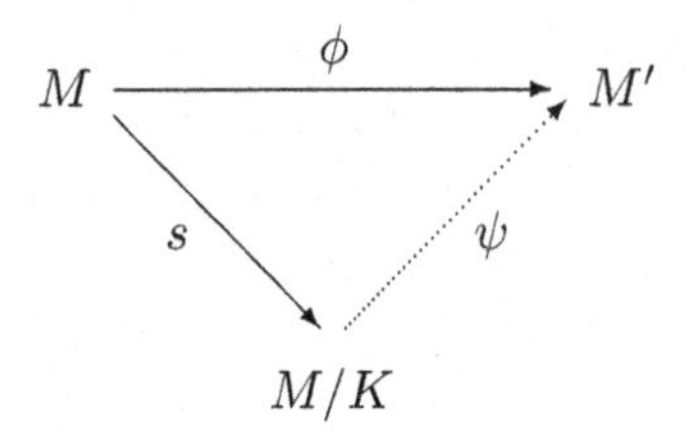

As we have seen, if V is a finite dimensional vector space over a field K and W is a subspace of V, then W is finite dimensional with $\dim(W) \leq \dim(V)$ (Proposition 3.5). If we generalize to arbitrary rings, we cannot hope for the proposition to remain true (can you think of an example where it fails?) However, in the case that R is a PID we have the following.

Proposition 3.14. *Let R be a PID, let F be a free R-module of rank m, and let M be a submodule of F. Then M is a free R-module of rank $l \leq m$.*

Proof. Our proof is by induction on m. For the trivial case, assume that $m = 1$, and let $\{x_1\}$ be an R-basis for F. Let M be an R-submodule of F. Let I be defined as

$$I = \{r \in R : rx_1 \in M\}.$$

Then I is an ideal of R, and since R is a PID, $I = Ra$ for some $a \in R$. If $a = 0$, then $M = 0$, and consequently, M has rank 0. If $a \neq 0$, then $M = R(ax_1)$, and so, M has rank 1.

For the induction hypothesis, we assume that every submodule of a free R-module of rank $m-1$ is free of rank $l \leq m-1$. Let F be a free R-module of rank m on the basis $\{x_1, x_2, \ldots, x_m\}$. Let M be a submodule of F.

Let

$$I = \{r \in R : x = r_1x_1 + r_2x_2 + \cdots + r_{m-1}x_{m-1} + rx_m \in M,$$
$$\text{for some } r_1, r_2, \ldots, r_{m-1} \in R\}.$$

Then I is an ideal of R so that $I = Ra$ for some $a \in R$. If $a = 0$, then M is a submodule of the free rank $m-1$ R-module $Rx_1 \oplus \cdots \oplus Rx_{m-1}$ and so by the induction hypothesis, M is free of rank $\leq m-1$. If $a \neq 0$, put

$$N = M \cap (Rx_1 \oplus Rx_2 \oplus \cdots \oplus Rx_{m-1}).$$

Then N is a submodule of the free module $Rx_1 \oplus \cdots \oplus Rx_{m-1}$, and by the induction hypothesis, N is free of rank $\leq m-1$. Let w be an element of M of the form

$$w = r_1x_1 + r_2x_2 + \cdots + r_{m-1}x_{m-1} + ax_m,$$

for some $r_1, r_2, \ldots, r_{m-1} \in R$, and let

$$x = s_1x_1 + s_2x_2 + \cdots + s_{m-1}x_{m-1} + rx_m,$$

for $s_1, s_2, \ldots, s_{m-1}, r \in R$, be an element of M. Then there exists $c \in R$ for which $x - cw \in N$, thus

$$M = N + Rw.$$

Now if $w \in N$, then $M = N$ and so M is free of rank $l \leq m-1 < m$. If $w \notin N$, then $N \cap Rw = \{0\}$, and so M is isomorphic to the direct sum of R-modules

$$R \oplus R \oplus \cdots \oplus R$$

where the number of summands l satisfies $l \leq m$. We conclude that a submodule M of a free R-module of rank m is free of rank $l \leq m$. □

We can generalize Proposition 3.14 to Noetherian rings.

Proposition 3.15. *Let R be a Noetherian ring, let $\bigoplus_{i=1}^m R$ denote the direct sum of m copies of R, and let M be an R-submodule of $\bigoplus_{i=1}^m R$. Then M is finitely generated.*

Proof. Our proof again uses induction on m. For the trivial case, assume that $m = 1$. Let M be an R-submodule of R. Then M is an ideal I of R, and since R is Noetherian, I is finitely generated.

For the induction hypothesis, we assume that every R-submodule of a direct sum $\bigoplus_{i=1}^{m-1} R$ of $m-1$ copies of R is finitely generated.

Let M be an R-submodule of $\bigoplus_{i=1}^m R$. Let

$$I = \{r \in R : x = (r_1, r_2, \ldots, r_{m-1}, r) \in M,$$
$$\text{for some } r_1, r_2, \ldots, r_{m-1} \in R\}.$$

Then I is an ideal of R, hence $I = Ra_1 + Ra_2 + \cdots + Ra_k$ for some $a_1, a_2, \ldots, a_k \in R$. If $I = 0$, then M is a submodule of $\bigoplus_{i=1}^{m-1} R$, and so by the induction hypothesis, M is finitely generated. If $I \neq 0$, put

$$N = M \cap \bigoplus_{i=1}^{m-1} R.$$

Then N is a submodule of $\bigoplus_{i=1}^{m-1} R$, and by the induction hypothesis, N is finitely generated. Let $w_1, w_2, \ldots, w_k$ be an elements of M of the form

$$w_i = (r_{1,i}, r_{2,i}, \ldots, r_{m-1,i}, a_i),$$

for some $r_{1,i}, r_{2,i}, \ldots, r_{m-1,i} \in R$, $1 \le i \le k$, and let

$$x = (s_1, s_2, \ldots, s_{m-1}, r),$$

for $s_1, s_2, \ldots, s_{m-1}, r \in R$, be an element of M. Then there exist $c_1, c_2, \ldots, c_k \in R$ for which

$$x - (c_1 w_1 + c_2 w_2 + \cdots + c_k w_k) \in N.$$

Thus

$$M = N + Rw_1 + Rw_2 + \cdots + Rw_k,$$

and so M is finitely generated. □

Corollary 3.1. *Let R be a Noetherian ring, let M be a finitely generated module over R and let N be a submodule of M. Then N is finitely generated.*

Proof. Let $a_1, a_2, \ldots, a_m$ be a generating set for M. Let $\bigoplus_{i=1}^{m} R$ denote the direct sum of m copies of R. Then there is a surjective R-module homomorphism

$$\phi : \bigoplus_{i=1}^{m} R \to M$$

defined as $\phi(r_1, r_2, \ldots, r_m) = r_1 a_1 + r_2 a_2 + \cdots + r_m a_m$. Now $\phi^{-1}(N)$ is an R-submodule of $\bigoplus_{i=1}^{m} R$ which is finitely generated by Proposition 3.15. Let $\{s_1, s_2, \ldots, s_k\}$ denote a generating set for $\phi^{-1}(N)$. Then

$$\{\phi(s_1), \phi(s_2), \ldots, \phi(s_k)\}$$

is a generating set for N. □

3.3 Projective Modules

In this section we introduce the notion of a short exact sequence of R-modules, which extends the concept of short exact sequence of groups that we discussed in §1.3. We specialize to split short exact sequences and define the notion of a projective R-module. Projective modules are generalizations of free R-modules (and thus generalizations of vector spaces). If M is a finitely generated and projective module over R, then M is a direct summand of a free R-module of finite rank. Moreover, if R is also local, then M is free of finite rank.

* * *

A sequence of homomorphisms and modules of the form

$$0 \xrightarrow{\gamma_0} M_1 \xrightarrow{\gamma_1} M_2 \xrightarrow{\gamma_2} M_3 \xrightarrow{\gamma_3} 0 \tag{3.1}$$

in which $\mathrm{im}(\gamma_i) = \ker(\gamma_{i+1})$ for $i = 0, 1, 2$ is a **short exact sequence of homomorphisms and modules**. The short exact sequence (3.1) is **split** if there exists an R-module map $s : M_3 \to M_2$ with $\gamma_2 s = \mathrm{id}_{M_3}$, where $\mathrm{id}_{M_3} : M_3 \to M_3$ is the identity map.

Example 3.1. Let $R = Z$ and let M be a Z-module in which there exist elements m_1, m_2, m_3 that satisfy the relations $3m_1 - m_2 = 0$ and $-m_1 - m_3 = 0$, only. Let T denote the Z-submodule of M generated by m_1, m_2, m_3. Then by Proposition 3.12, T is a quotient of a free Z-module F of finite rank; F is the free Z-module on the basis $\{x_1, x_2, x_3\}$. There exists a surjective Z-module homomorphism $\gamma_2 : F \to T$, given as $x_i \mapsto m_i$, for $i = 1, 2, 3$, and $F/N \cong T$ with $N = \ker(\gamma_2)$.

By Proposition 3.14, N is free over Z of rank ≤ 3, in fact, N is a free rank 2 Z-module on the basis $\{3x_1 - x_2, -x_1 - x_3\}$. Let $\gamma_1 : N \to F$ be the inclusion. Then there exists a short exact sequence of Z-modules

$$0 \xrightarrow{\gamma_0} N \xrightarrow{\gamma_1} F \xrightarrow{\gamma_2} T \xrightarrow{\gamma_3} 0$$

with γ_0, γ_3 the obvious maps. This short exact sequence is split since the map $s : T \to F$ defined by $s(m_1) = x_1$, $s(m_2) = 3x_1$, $s(m_3) = -x_1$ induces a Z-module homomorphism with $\gamma_2 s = \mathrm{id}_T$.

Example 3.2. Let $\phi : Z_6 \to Z_3$ be the ring homomorphism defined as $\phi(m) = m \bmod 3$. Then Z_3 is a Z_6-module with scalar multiplication defined as $m \cdot n = \phi(m)n$. As a Z_6-module, Z_3 is generated by $\{1\}$. By

Proposition 3.12, Z_3 is the quotient of a free rank one Z_6 module, indeed $Z_6/\{0,3\} \cong Z_3$. One has the short exact sequence of Z_6-modules

$$0 \xrightarrow{\gamma_0} \{0,3\} \xrightarrow{\gamma_1} Z_6 \xrightarrow{\phi} Z_3 \xrightarrow{\gamma_3} 0$$

with γ_0, γ_3 the obvious maps. Now, $s : Z_3 \to Z_6$ defined as $s(0) = 0$, $s(1) = 4$, $s(2) = 2$ is a Z_6-module homomorphism that satisfies $\phi s = \mathrm{id}_{Z_3}$. Thus the short exact sequence is split.

Definition 3.11. An R-module P is **projective** if every short exact sequence

$$0 \longrightarrow A \longrightarrow B \xrightarrow{\phi} P \longrightarrow 0$$

is split.

Proposition 3.16. *An R-module M is projective if and only if there exist elements $\{a_\eta\}_{\eta\in J}$ of M, and R-module homomorphisms $\{\beta_\eta\}$, $\beta_\eta : M \to R$, $\eta \in J$ that satisfy*

(i) for each $x \in M$, $\beta_\eta(x) = 0$ for all but a finite number of η,

(ii) for each $x \in M$, $x = \sum_{\eta\in J} \beta_\eta(x)a_\eta$.

Proof. We prove the "if" part and leave the converse for an exercise.

Observe that (ii) implies that $\{a_\eta\}_{\eta\in J}$ is a generating set for M. Let F be the free R-module on the set of indeterminates $\{x_\eta\}_{\eta\in J}$. Then the R-module homomorphism $\phi : F \to M$ defined by $\phi(x_\eta) = a_\eta$ is surjective. Define a map $s : M \to F$ by the rule $s(m) = \sum_{\eta\in J} \beta_\eta(m)x_\eta$. Then s is a homomorphism of R-modules with $\phi s = \mathrm{id}_M$, and thus the short exact sequence

$$0 \longrightarrow \ker(\phi) \longrightarrow F \xrightarrow{\phi} M \longrightarrow 0$$

is split. Now let

$$0 \longrightarrow A \longrightarrow B \xrightarrow{\psi} M \longrightarrow 0$$

be any short exact sequence. Let $g : F \to B$ be the map defined as $g(x_\eta) = a_\eta$ where a_η is so that $\psi(a_\eta) = \phi(x_\eta)$ for all $\eta \in J$. Then g is an R-module map for which $\phi = \psi g$, cf. [Rotman (2002), Theorem 7.52]. It follows that $gs : M \to B$ is an R-module map for which $\psi gs = \mathrm{id}_M$, and thus M is projective. □

Proposition 3.16 implies that any free R-module is projective. Proof: Let $\{a_\eta\}_{\eta \in J}$ be an R-basis for M and set $\beta_\gamma(a_\eta) = \delta_{\gamma,\eta}$. Consequently, the Z-module T of Example 3.1 is projective since T is free rank one on the basis $\{m_1\}$.

Regarding the Z_6-module Z_3 of Example 3.2, one sees that the element $\{1\}$ of Z_3, together with Z_6-module map $s : Z_3 \to Z_6$ satisfy conditions (i) and (ii) of Proposition 3.16. Consequently, Z_3 is a projective Z_6-module. Z_3 is not free over Z_6, however, and so here we have an example of a projective module that is not free.

Proposition 3.17. *Suppose N is a submodule of M. If the quotient module M/N is projective then there exists a submodule A of M with $A \cong M/N$ and $M = N \oplus A$.*

Proof. Let $\phi : M \to M/N$ denote the canonical surjection. There is a short exact sequence of R-modules

$$0 \longrightarrow N \longrightarrow M \xrightarrow{\phi} M/N \longrightarrow 0$$

and so, since M/N is projective, there exists an R-module map $s : M/N \to M$ with $\phi s = Id_{M/N}$. We show that $M \cong N \oplus M/N$. Let $m \in M$. Then

$$\phi(m - s(\phi(m))) = \phi(m) - \phi s(\phi(m)) = \phi(m) - \phi(m) = 0,$$

thus $m = n + s(\phi(m))$ for some $n \in N$. This says that $M \subseteq N + s(M/N)$. Now since $N + s(M/N) \subseteq M$, we conclude that $M = N + s(M/N)$.

Now suppose that $m = n + r$ for $n \in N$, $r \in s(M/N)$ with $r = s(t+N)$, $t + N \in M/N$. Then $\phi(m) = \phi(r) = t + N$, and so $t + N$ is uniquely determined by m, and consequently, r and n are uniquely determined by m. Thus $M = N \oplus s(M/N)$. Now since $A = s(M/N) \cong M/N$, the result follows. □

Corollary 3.2. *Every finitely generated and projective R-module is a direct summand of a free module of finite rank.*

Proof. Exercise. □

If R is a local ring, then a finitely generated and projective R-module is free over R. To prove this result we shall use the following special case of Nakayama's Lemma.

Proposition 3.18. *Let R be a local ring with maximal ideal m. Let M be a finitely generated R-module.*

(i) If $mM = M$, then $M = 0$,

(ii) Let $M \to M/mM$ denote the canonical surjection of R-modules. Let $\{x_1, x_2, \ldots, x_k\}$ be a subset of M, and suppose that the set of images $\{\overline{x}_i\}$ generates the quotient M/mM as an R-module. Then the set $\{x_i\}$ generates M.

Proof. To prove (i), assume that $M \neq 0$, and let $S = \{x_1, x_2, \ldots, x_k\}$ be a minimal generating set for M. Since $mM = M$, there exists elements $r_1, r_2, \ldots, r_k \in m$ for which $x_1 = \sum_{i=1}^{k} r_i x_i$. Thus,

$$(1 - r_1)x_1 = x_1 - r_1 x_1 = \sum_{i=2}^{k} r_i x_i.$$

Since $r_1 \in m$, $1 - r_1 \notin m$, and so, $1 - r_1$ is a unit of R. (Proof: if $1 - r_1$ is not a unit, then $1 - r_1$ is contained in some maximal ideal J of R by Proposition 2.23. But since R is local, $J = m$.) Now,

$$x_1 = \sum_{i=2}^{k} (1 - r_1)^{-1} r_i x_i,$$

and so, $S \backslash \{x_1\}$ is a generating set for M which contradicts the minimality of S.

For (ii), suppose $S = \{x_1, x_2, \ldots, x_k\}$ is a subset of M for which $\{\overline{x}_1, \overline{x}_2, \ldots, \overline{x}_k\}$ generates M/mM. Let B be the submodule of M generated by S, that is, B consists of all sums of the form $\sum_{i=1}^{k} r_i x_i$ for $r_i \in R$. Then $(M/B)/m(M/B) = M/(mM + B) = 0$, since $mM + B = M$. Thus, $(M/B) = m(M/B)$. Now by (i), $M/B = 0$ which yields $M = B$, and so, $\{x_i\}$ generates M. □

Proposition 3.19. *Let R be a local ring with maximal ideal m. Let M be a finitely generated and projective R-module. Then M is a free R-module.*

Proof. Assume $M \neq 0$, and let $S = \{x_1, x_2, x_3, \ldots, x_k\}$ be a minimal set of generators for M. The quotient module M/mM is a module over R and also over R/m, which is a field. Let $\overline{x}_i$ denote the images of the x_i under the canonical surjection $M \to M/mM$. Then the elements $\{\overline{x}_i\}$ generate the R-module M/mM. We claim that $\overline{x}_i \notin mM$ for all i. Otherwise, if some $\overline{x}_{i'} \in mM$, then the set

$$\{\overline{x}_1, \overline{x}_2, \ldots, \overline{x}_k\} \backslash \{\overline{x}_{i'}\}$$

generates M/mM. Now by Proposition 3.18(ii), the set $S\backslash\{x_{i'}\}$ generates M, contradicting the minimality of S.

We next claim that $\{\overline{x}_i\}$ is a basis for the R/m-module M/mM. To this end, suppose that

$$\overline{r}_1\overline{x}_1 + \overline{r}_2\overline{x}_2 + \cdots + \overline{r}_k\overline{x}_k \in mM,$$

for elements $\overline{r}_i \in R/m$. Suppose that $r_{i'} \notin m$ for some i'. Then

$$\overline{r}_{i'}\overline{x}_{i'} \in (\sum_{\substack{i=1,\\ i\neq i'}}^{k} -\overline{r}_i\overline{x}_i) + mM.$$

Thus,

$$\overline{x}_{i'} \in (\sum_{\substack{i=1,\\ i\neq i'}}^{k} -(\overline{r}_{i'})^{-1}\overline{r}_i\overline{x}_i) + mM,$$

and so, $\{\overline{x}_1, \overline{x}_2, \ldots, \overline{x}_k\}\backslash\{\overline{x}_{i'}\}$ generates M/mM, again contradicting the minimality of S (by Proposition 3.18(ii)). Thus $r_i \in m$ for all i and so, $\{\overline{x}_i\}$ is a basis for M/mM.

Now let R^k denote the free R-module of rank k with basis $\{b_i\}$, and let $\phi : R^k \to M$ be the surjective map of R-modules defined by $\phi(b_i) = x_i$. Then there is a short exact sequence of R-modules

$$0 \longrightarrow \ker(\phi) \longrightarrow R^k \longrightarrow M \longrightarrow 0,$$

which is split since M is projective. By Proposition 3.17, $R^k = M' \oplus \ker(\phi)$ where $M' \cong M$. Now ϕ induces an isomorphism of vector spaces $R^k/mR^k \to M/mM$, and so, $\ker(\phi) \subseteq mR^k$. Moreover, $mR^k \cong mM \oplus m\ker(\phi)$, and so $\ker(\phi)$ injects into $mM \oplus m\ker(\phi)$. It follows that $\ker(\phi) \subseteq m\ker(\phi)$, and thus $\ker(\phi) = m\ker(\phi)$. Therefore, by Proposition 3.18(i), $\ker(\phi) = 0$. Thus ϕ is an isomorphism of R-modules, and consequently, M is free over R. □

Let R be an integral domain, let M be an R-module and let P be a prime ideal of R. We can often say something globally about M if we localize at P. This is possible in view of the following proposition.

Let R_P denote the localization of R at P and set

$$M_P = \{m/s : \ m \in M, s \in S = R\backslash P\}.$$

Then M_P is an R_P-module, with R_P a local ring (Proposition 2.47).

Proposition 3.20. *With notation as above, and with the identification $M = M/1$, one has* $M = \bigcap_{P \text{ prime}} M_P$.

Proof. One easily obtains $M \subseteq \bigcap_{P \text{ prime}} M_P$.

Let $x \in \bigcap_{P \text{ prime}} M_P$ and consider the ideal of R defined as

$$I = \{a \in R : \ ax \in M\}.$$

Now $x = m/s$ with $m \in M$, and $s \in R$ with $s \notin P$ for all primes P in R. Since $s \in I$, I is not contained in any prime ideal of R. Thus by Corollary 2.3, $I = R$, and so $1x = x \in M$. □

This formula underscores that a module can be locally free yet not free: the R_P-basis B that could exist for each M_P may not be in the intersection of all of the M_P, and hence may not yield an R-basis for M.

As an example, we recall the illustration of Proposition 3.12; we let $R = Z[(1 + \sqrt{-23})/2]$ (this is the ring of integers of the field extension $\mathbb{Q}(\sqrt{-23})$). Let $M = 3R + (2 + \sqrt{-23})R$. By [Rot02, Proposition 11.98], M is a finitely generated and projective R-module and so by Proposition 3.19, M_P is a free R_P-module for each prime ideal P of R. However, (as we will show in Chapter 4) M is not a principal ideal of R, and hence M is not a free R-module.

Of course, if the R-module M is R-free with basis B, then $B \subseteq \bigcap M_P$ and so the image of B in each M_P will be an R_P-basis for M_P.

3.4 Tensor Products

In this section we introduce bilinear maps and show how they can be used to construct the tensor product $M \otimes_R N$ of R-modules M, N. We specialize to the tensor product $V \otimes_K W$ of two vector spaces over K and relate the linear dual $(V \otimes_K W)^*$ to the dual spaces V^* and W^*.

* * *

Let R be a commutative ring with unity and let M, N, A be R-modules.

Definition 3.12. A function $f : M \times N \to A$ is **R-bilinear** if for all $a, a' \in M$, $b, b' \in N$, $r \in R$,

(i) $f(a + a', b) = f(a, b) + f(a', b)$,

(ii) $f(a, b + b') = f(a, b) + f(a, b')$,

(iii) $f(ra, b) = rf(a, b) = f(a, rb)$.

Definition 3.13. A **tensor product** of M, N over R is an R-module $M \otimes_R N$ together with an R-bilinear map

$$f : M \times N \to M \otimes_R N$$

so that for every R-module A and every R-bilinear map $h : M \times N \to A$ there exists a unique module homomorphism $\tilde{h} : M \otimes_R N \to A$ for which $\tilde{h}f = h$, that is, the following diagram commutes.

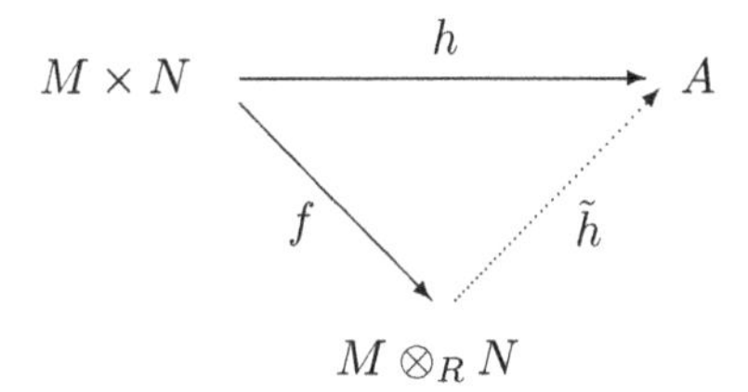

We construct a tensor product as follows. Let $F\langle M \times N\rangle$ denote the free R module on the set $M \times N$. Let J be the submodule of $F\langle M \times N\rangle$ generated by the quantities

$$(a + a', b) - (a, b) - (a', b), \quad (a, b + b') - (a, b) - (a, b'),$$

$$r(a, b) - (ra, b), \quad r(a, b) - (a, rb),$$

for all $a, a' \in M$, $b, b' \in N$, $r \in R$. Let $\iota : M \times N \to F\langle M \times N\rangle$ denote the natural inclusion and let $s : F\langle M \times N\rangle \to F\langle M \times N\rangle/J$ be the canonical surjection. Let $f = s\iota$. We show that the quotient module $F\langle M \times N\rangle/J$ together with f is a tensor product $M \otimes_R N$ by showing that they "solve a universal mapping problem", that is, we show that $F\langle M \times N\rangle/J$ together with f satisfies the conditions of Definition 3.13.

Proposition 3.21. *$F\langle M \times N\rangle/J$ together with f is a tensor product of M, N over R.*

Proof. We show that the conditions of Definition 3.13 are satisfied. Let A be an R-module and let $h : M \times N \to A$ be a bilinear map. There exists an R-module homomorphism

$$\phi : F\langle M \times N \rangle \to A,$$

defined as $\phi(a,b) = h(a,b), \forall (a,b) \in M \times N$. Since ϕ is R-bilinear, $J \subseteq \ker(\phi)$, and so, by the UMPK, there exists an R-module homomorphism

$$\tilde{h} : F\langle M \times N \rangle / J \to A,$$

defined as $\tilde{h}((a,b)+J) = \phi(a,b) = h(a,b)$. Now for all $(a,b) \in M \times N$, $\tilde{h}s\iota(a,b) = h(a,b)$, and so, $\tilde{h}f = h$. Moreover, $\tilde{h}$ is unique because $\{(a,b) + J : (a,b) \in M \times N\}$ is a set of generators for $F\langle M \times N \rangle / J$. □

As a consequence of Proposition 3.21, we write $F\langle M \times N \rangle / J = M \otimes_R N$, with the coset $(a,b)+J$ now written as the **tensor** $a \otimes b$. In $M \otimes_R N$, one has the tensor relations:

$$(a + a') \otimes b = a \otimes b + a' \otimes b,$$

$$a \otimes (b + b') = a \otimes b + a \otimes b',$$

$$ra \otimes b = a \otimes rb,$$

for all $a, a' \in M$, $b, b' \in N$, $r \in R$. For example, let $M = N = R[x]$, and consider the tensor product $R[x] \otimes_R R[x]$. In $R[x] \otimes_R R[x]$, $(x+1) \otimes x^2 = x \otimes x^2 + 1 \otimes x^2$, and $2x \otimes x = x \otimes 2x$. If M is any R-module, then $R \otimes_R M \cong M$, as R-modules, the isomorphism being $r \otimes s \mapsto rs$.

Proposition 3.22. *Let M_1, M_2 be R-modules and let N_1 be an R-submodule of M_1 and let N_2 be an R-submodule of M_2. Then there is an isomorphism of R-modules*

$$M_1/N_1 \otimes_R M_2/N_2 \cong (M_1 \otimes_R M_2)/(N_1 \otimes_R M_2 + M_1 \otimes_R N_2).$$

Proof. First note that there is an R-bilinear map

$$h : M_1 \times M_2 \to M_1/N_1 \otimes_R M_2/N_2,$$

defined by $h(a,b) = (a+N_1) \otimes (b+N_2)$. Since $M_1 \otimes_R M_2$ is a tensor product there exists an R-module map

$$\tilde{h} : M_1 \otimes_R M_2 \to M_1/N_1 \otimes_R M_2/N_2,$$

defined as $\tilde{h}(a \otimes b) = (a+N_1) \otimes (b+N_2)$. Now $N_1 \otimes_R M_2 + M_1 \otimes_R N_2 \subseteq \ker(\tilde{h})$, and so by Proposition 3.13 (UMPK), there exists an R-module map

$$\alpha : (M_1 \otimes_R M_2)/(N_1 \otimes_R M_2 + M_1 \otimes_R N_2) \to M_1/N_1 \otimes_R M_2/N_2,$$

with $\alpha(a \otimes b + (N_1 \otimes_R M_2 + M_1 \otimes_R N_2)) = (a + N_1) \otimes (b + N_2)$.

Next, let

$$l : M_1/N_1 \times M_2/N_2 \to (M_1 \otimes_R M_2)/(N_1 \otimes_R M_2 + M_1 \otimes_R N_2)$$

be the relation defined as

$$l(a + N_1, b + N_2) = a \otimes b + (N_1 \otimes_R M_2 + M_1 \otimes_R N_2),$$

for $a \in M_1$, $b \in M_2$. We claim that l is actually a function on $M_1/N_1 \times M_2/N_2$. To this end, let $x = a + n$, $y = b + n'$ for some $n \in N_1$, $n' \in N_2$. Then

$$\begin{aligned} l(x + N_1, y + N_2) &= x \otimes y + (N_1 \otimes_R M_2 + M_1 \otimes_R N_2) \\ &= (a + n) \otimes (b + n') + (N_1 \otimes_R M_2 + M_1 \otimes_R N_2) \\ &= a \otimes b + a \otimes n' + n \otimes b + n \otimes n' \\ &\quad + (N_1 \otimes_R M_2 + M_1 \otimes_R N_2), \end{aligned}$$

and thus $l(a + N_1, b + N_2) - l(x + N_1, y + N_2) = N_1 \otimes_R M_2 + M_1 \otimes_R N_2$. It follows that l is a well-defined function on $M_1/N_1 \times M_2/N_2$. As one can easily check, l is R-bilinear and since $M_1/N_1 \otimes_R M_2/N_2$ is a tensor product, there exists an R-module map

$$\tilde{l} : M_1/N_1 \otimes_R M_2/N_2 \to (M_1 \otimes_R M_2)/(N_1 \otimes_R M_2 + M_1 \otimes_R N_2),$$

defined as $\tilde{l}((a + N_1) \otimes (b + N_2)) = a \otimes b + (N_1 \otimes_R M_2 + M_1 \otimes_R N_2)$. Clearly $\alpha\tilde{l}$ is the identity on $M_1/N_1 \otimes M_2/N_2$ and $\tilde{l}\alpha$ is the identity on $(M_1 \otimes_R M_2)/(N_1 \otimes_R M_2 + M_1 \otimes_R N_2)$, thus $\tilde{l}$ is an isomorphism. □

Let V, W be vector spaces over a field K. Then the tensor product $V \otimes W$ is a K-vector space, as are $V^* \otimes W^*$ and $(V \otimes W)^*$.

Proposition 3.23. *Let V, W be finite dimensional vector spaces over a field K with* $\dim(V) = m$, $\dim(W) = n$. *Then $V \otimes_K W$ is finite dimensional over K with $dim(V \otimes W) = mn$. Moreover, if $\{b_i\}_{i=1}^m$ is a K-basis for V and $\{c_j\}_{j=1}^n$ is a K-basis for W, then $B = \{b_i \otimes c_j\}$,* $1 \le i \le m$, $1 \le j \le m$, *is a K-basis for $V \otimes_K W$.*

Proof. Certainly the submodule of $V \otimes W$ generated by $B = \{b_i \otimes c_j\}$, $1 \leq i \leq m$, $1 \leq j \leq m$ is all of $V \otimes_K W$. So we need to show that B is a linearly independent subset of $V \otimes_K W$. Suppose that

$$\sum_{i=1}^{m}\sum_{j=1}^{n} r_{i,j}(b_i \otimes c_j) = 0.$$

If $r_{i,j} \neq 0$ for some i, j, say i^*, j^*, then

$$b_{i^*} \otimes c_{j^*} + \sum_{\substack{i=1,\\ i\neq i^*}}^{m} \sum_{\substack{j=1,\\ j\neq j^*}}^{n} \frac{r_{i,j}}{r_{i^*,j^*}}(b_i \otimes c_j) = 0. \tag{3.2}$$

Thus the quantity (3.2) is in the subspace of the free K-module $F\langle V \times W\rangle$ generated by the quantities

$$(a + a', b) - (a, b) - (a', b), \quad (a, b + b') - (a, b) - (a, b'),$$
$$r(a, b) - (ra, b), \quad r(a, b) - (a, rb),$$

for all $a, a' \in V$, $b, b' \in W$, $r \in K$. But this is impossible since $\{b_i\}_{i=1}^{m}$ is a linearly independent subset of V. It follows that $r_{i,j} = 0$ for all i, j. □

Proposition 3.24. $V^* \otimes W^* \subseteq (V \otimes W)^*$.

Proof. Let $f \in V^*$, $g \in W^*$ and let $f \times g : V \times W \to K$ be the map defined by

$$(f \times g)(a, b) = f(a)g(b)$$

for all $a \in V$, $b \in W$. As one can easily check, $f \times g$ is K-bilinear. For instance,

$$\begin{aligned}(f \times g)(a + a', b) &= f(a + a')g(b)\\ &= (f(a) + f(a'))g(b)\\ &= f(a)g(b) + f(a')g(b)\\ &= (f \times g)(a, b) + (f \times g)(a', b),\end{aligned}$$

and so, $f \times g$ is linear in the first component.

Now, since $V \otimes W$ is a tensor product over K, there exists a unique linear transformation $\tilde{h} : V \otimes W \to K$ defined as $\tilde{h}(a \otimes b) = (f \times g)(a, b) = f(a)g(b)$. Set $f \otimes g = \tilde{h}$. This shows the containment $V^* \otimes W^* \subseteq (V \otimes W)^*$. □

When do we have equality in Proposition 3.24?

Proposition 3.25. *Suppose that V, W are finite dimensional vector spaces over K. Then*

$$V^* \otimes W^* = (V \otimes W)^*.$$

Proof. Let $\dim(V) = m$ and $\dim(W) = n$. Let $\{b_1, b_2, \ldots, b_m\}$ be a basis for V, and let $\{f_1, f_2, \ldots, f_m\}$ denote the corresponding dual basis for V^* with $f_i(b_j) = \delta_{i,j}$. Let $\{c_1, c_2, \ldots, c_n\}$ be a basis for W, and let $\{g_1, g_2, \ldots, g_n\}$ denote the corresponding dual basis for W^* with $g_i(c_j) = \delta_{i,j}$. By Proposition 3.23, the set $\{b_i \otimes c_j\}$, $1 \leq i \leq m$, $1 \leq j \leq n$, is a basis for $V \otimes W$. By Proposition 3.7, the set $\{\psi_{\alpha,\beta}\}$, $1 \leq \alpha \leq m$, $1 \leq \beta \leq n$, with

$$\psi_{\alpha,\beta}(b_i \otimes c_j) = \delta_{\alpha,i}\delta_{\beta,j},$$

is a basis for $(V \otimes W)^*$.

By Proposition 3.24 for each pair α, β, $1 \leq \alpha \leq m$, $1 \leq \beta \leq n$, there is a K-linear map

$$f_\alpha \otimes g_\beta : V \otimes W \to K,$$

defined by

$$(f_\alpha \otimes g_\beta)(b_i \otimes c_j) = f_\alpha(b_i)g_\beta(c_j).$$

Clearly, $f_\alpha \otimes g_\beta = \psi_{\alpha,\beta}$, for all α, β, and so $V^* \otimes W^* = (V \otimes W)^*$. $\square$

If V and W are infinite dimensional, however, then $V^* \otimes W^*$ is a proper subset of $(V \otimes W)^*$.

Proposition 3.26. *Suppose that V, W are infinite dimensional vector spaces over K. Then $V^* \otimes W^* \subset (V \otimes W)^*$.*

Proof. Any element $\sum f \otimes g \in V^* \otimes W^*$ determines an element of $(V \otimes W)^*$ defined as

$$(\sum f \otimes g)(\sum a \otimes b) = \sum\sum f(a)g(b),$$

hence $V^* \otimes W^* \subseteq (V \otimes W)^*$. To show that the containment is proper, we proceed as follows.

By Proposition 3.2 there exists a K-basis $\{b_\alpha\}_{\alpha \in I}$ for V and a K-basis $\{c_\beta\}_{\beta \in J}$ for W. Let $\{f_\alpha\}_{\alpha \in I}$ be a collection of maps $f_\alpha : V \to K$ defined by

$$f_\alpha(\sum_{\beta \in I} r_\beta b_\beta) = r_\alpha, \quad r_\beta \in K,$$

where $r_\beta = 0$ for all but a finite number of β. Similarly, let $\{g_\beta\}_{\beta\in J}$ be a collection of maps $g_\beta : W \to K$ defined by

$$g_\beta(\sum_{\alpha\in J} s_\alpha c_\alpha) = s_\beta, \quad s_\alpha \in K,$$

where $s_\alpha = 0$ for all but a finite number of α. Then $\{f_\alpha\} \subseteq V^*$, $\{g_\beta\} \subseteq W^*$. There is a countably infinite subcollection $\{f_{\alpha_i}\}_{i=1}^\infty \subseteq \{f_\alpha\}$ and a countably infinite subcollection $\{g_{\beta_j}\}_{j=1}^\infty \subseteq \{g_\beta\}$. Now $\sum_{i=1}^\infty f_{\alpha_i} \otimes g_{\beta_i}$ is an infinite sum that becomes a finite sum upon evaluation at an element of $V \otimes W$. Thus

$$\sum_{i=1}^\infty f_{\alpha_i} \otimes g_{\beta_i} \in (V \otimes W)^* \backslash (V^* \otimes W^*).$$

□

As an example of this phenomenon we take $V = W = K[x]$, the vector space of polynomials over a field K. Then $\{1, x, x^2, x^3, \ldots\}$ is a basis for $K[x]$. Let $\{e_0, e_1, e_2, e_3, \ldots\}$ be a collection of linear functionals in $K[x]^*$ defined as

$$e_i(x_j) = \delta_{i,j},$$

for all i, j. Note that $\{e_i\}_{i=1}^\infty$ is a linearly independent subset of $K[x]^*$, yet it is not a basis for $K[x]^*$ over K. As one can easily check, $\sum_{i=1}^\infty e_i \otimes e_i$ is an element of $(K[x] \otimes K[x])^*$ that is not in $K[x]^* \otimes K[x]^*$.

Every element $f \in K[x]^*$ has a unique representation as an infinite sum

$$f = \sum_{n=0}^\infty s_n e_n = s_0 e_0 + s_1 e_1 + s_2 e_2 + \cdots,$$

for $s_n \in K$, $n \geq 0$. Thus elements of $K[x]^*$ can be identified with sequences $\{s_n\}_{n=0}^\infty$ of elements in K.

3.5 Algebras

In this section we consider algebras over a commutative ring with unity R. An R-algebra A is an object that is both a ring A and an R-module – the scalar multiplication is given through a ring homomorphism $\lambda : R \to A$. We specialize to commutative R-algebras, and show that every finitely generated R-algebra is the quotient of a polynomial ring. We compute all of the R-algebra homomorphisms $A \to B$, where A is finitely generated.

* * *

Definition 3.14. Let R be a commutative ring with unity, 1_R and let A be a ring with unity 1_A. Then A is an **algebra over** R if there exists a homomorphism of rings $\lambda : R \to A$ with unity for which $\lambda(R)$ is contained in the center of A, that is, $\lambda(r)a = a\lambda(r)$ for all $r \in R$, $a \in A$. Then A is an R-module with scalar multiplication $R \times A \to A$ given as $ra = \lambda(r)a = a\lambda(r)$, for all $r \in R$, $a \in A$. The map λ is the **R-algebra structure map of** A, which we sometimes denote as λ_A. Specializing to $a = 1_A$ yields $\lambda(r) = r1_A$, and so $\lambda(R)$ is a subring of A consisting of all R-scalar multiples of 1_A. If λ is 1-1 (for instance, if R is a field), then λ sends R to an isomorphic copy of R in A; we can then identify R with $\lambda(R)$, and so $\lambda(r) = r$, for $r \in R$.

An R-algebra A is **commutative** if A is a commutative ring with unity.

For example, the polynomial ring $R[x]$ is a commutative R-algebra where the structure map $\lambda : R \to R[x]$ is given by $\lambda(r) = r$ (λ is 1-1; $r \neq 0$ is a degree 0 polynomial; $r = 0$ has degree $-\infty$). The monoid ring RS is an R-algebra with R-algebra structure map $\lambda : R \to RS$ defined as $\lambda(r) = r1 = r$, for all $r \in R$ (again λ is 1-1). Likewise, the group ring RG for G a finite group, is an R algebra with structure map $\lambda : R \to RG$, $r \mapsto r$.

For another example, let A be an R-algebra, and let J be an ideal of A (as a ring). Then the quotient ring A/J is an R-algebra. Indeed, if λ_A is the R-algebra structure map of A, then the R-module structure of A/J is given by the ring homomorphism $\lambda : R \to A/J$ defined as $r \cdot (a + J) = \lambda_A(r)a + J$.

Let A and B be algebras over R and let $A \otimes_R B$ be the tensor product of A and B as R-modules. Then the R-module $A \otimes_R B$ can be endowed with the structure of an R-algebra. First note that $A \otimes_R B$ is a ring: addition is the module addition

$$(\sum_{i=1}^{k} a_i \otimes b_i) + (\sum_{j=1}^{l} c_j \otimes d_j) = \sum_{i=1}^{k} a_i \otimes b_i + \sum_{j=1}^{l} c_j \otimes d_j$$

and multiplication is given as

$$(\sum_{i=1}^{k} a_i \otimes b_i)(\sum_{j=1}^{l} c_j \otimes d_j) = \sum_{i=1}^{k} \sum_{j=1}^{l} a_i c_j \otimes b_i d_j,$$

for $a_i, c_j \in A$, $b_i, d_j \in B$. The unity in $A \otimes B$ is $1_A \otimes 1_B$.

Next, let $\lambda : R \to A \otimes_R B$ be the ring homomorphism defined as $\lambda(r) = \lambda_A(r) \otimes 1_B$, where λ_A is the R-algebra structure map of A. Then

$$(\sum a \otimes b)\lambda(r) = \lambda(r)(\sum a \otimes b),$$

for all $r \in R$, $\sum a \otimes b \in A \otimes_R B$. Now $A \otimes_R B$, together with scalar multiplication defined through λ is an R-algebra, called the **tensor product R-algebra**.

What are the maps between R-algebras?

Definition 3.15. Let A, B be R-algebras with R-algebra structure maps $\lambda_A : R \to A$, $\lambda_B : R \to B$. Then $\phi : A \to B$ is an **R-algebra homomorphism** if ϕ is a ring homomorphism and

$$\phi(\lambda_A(r)a) = \lambda_B(r)\phi(a), \tag{3.3}$$

for all $r \in R$, $a \in A$ (ϕ preserves scalar multiplication).

Condition (3.3) is equivalent to the condition

$$\phi(\lambda_A(r)) = \lambda_B(r). \tag{3.4}$$

(Proof: in the case $a = 1_A$ in (3.3), one has

$$\phi(\lambda_A(r)) = \phi(\lambda_A(r)1_A) = \lambda_B(r)\phi(1_A) = \lambda_B(r)1_B = \lambda_B(r),$$

for $r \in R$. Conversely, multiplying both sides of (3.4) by $\phi(a)$ yields (3.3).) An **R-algebra isomorphism** is an R-algebra homomorphism which is a bijection.

The collection of all R-algebra homomorphisms from A to B is denoted by $\mathrm{Hom}_{\text{R-alg}}(A, B)$.

The following proposition extends Proposition 3.22 to R-algebras.

Proposition 3.27. *Let A be an R-algebra and let J be any ideal of A. Then there is an isomorphism of R-algebras*

$$A/J \otimes_R A/J \cong (A \otimes_R A)/(J \otimes_R A + A \otimes_R J).$$

Proof. By Proposition 3.22 the map

$$\tilde{l} : A/J \otimes_R A/J \to (A \otimes_R A)/(J \otimes_R A + A \otimes_R J),$$

defined as $\tilde{l}((a+N)\otimes(b+N)) = a\otimes b+(J\otimes_R A+A\otimes_R J)$ is an isomorphism of R-modules. One then shows that $\tilde{l}$ is a homomorphism of R-algebras. □

For the remainder of this section, all algebras are assumed to be commutative.

Proposition 3.28. *Let A be an R-algebra, and let $A \otimes_R A$ be the tensor product R-algebra. Then the map $m : A \otimes_R A \to A$ defined as*

$$m(\sum_{i=1}^{k} a_i \otimes b_i) = \sum_{i=1}^{k} a_i b_i,$$

is a homomorphism of R-algebras.

Proof. Clearly, addition is preserved under m. We show that multiplication is preserved. Notice the need for A to be commutative:

$$\begin{aligned} m\left((\sum_{i=1}^{k} a_i \otimes b_i)(\sum_{j=1}^{l} c_j \otimes d_j) \right) &= m(\sum_{i=1}^{k}\sum_{j=1}^{l} a_i c_j \otimes b_i d_j) \\ &= \sum_{i=1}^{k}\sum_{j=1}^{l} a_i c_j b_i d_j \\ &= \sum_{i=1}^{k}\sum_{j=1}^{l} a_i b_i c_j d_j \\ &= \sum_{i=1}^{k} a_i b_i \sum_{j=1}^{l} c_j d_j \\ &= m(\sum_{i=1}^{j} a_i \otimes b_i) m(\sum_{j=1}^{l} c_j \otimes d_j). \end{aligned}$$

Finally, we show that m respects the R-algebra structures on $A \otimes_R A$ and A:

$$\begin{aligned} m(r(\sum_{i=1}^{k} a_i \otimes b_i)) &= m(\sum_{i=1}^{k} \lambda_A(r) a_i \otimes b_i) \\ &= \sum_{i=1}^{k} \lambda_A(r) a_i b_i \\ &= \lambda_A(r) \sum_{i=1}^{k} a_i b_i \\ &= \lambda(r) m(\sum_{i=1}^{k} a_i \otimes b_i) \\ &= r m(\sum_{i=1}^{k} a_i \otimes b_i). \end{aligned}$$

□

Definition 3.16. Let $R[\{x_\alpha\}]$ be the R-algebra of polynomials in the indeterminates $\{x_\alpha\}_{\alpha \in J}$. Let B be an R-algebra. Let $\{b_\alpha\}_{\alpha \in J}$ be a family of elements in B indexed by J and let $f(\{x_\alpha\})$ be a polynomial in $R[\{x_\alpha\}]$.

Then the **evaluation of** $f(\{x_\alpha\})$ **at** $\{b_\alpha\}$ is the element $f(\{b_\alpha\}) \in B$. The map

$$\phi_{\{b_\alpha\}} : R[\{x_\alpha\}] \to B$$

defined as $f(\{x_\alpha\}) \mapsto f(\{b_\alpha\})$ is a homomorphism of R-algebras called the **evaluation homomorphism.**

Definition 3.17. Let A be an R-algebra, let $S = \{b_\alpha\}_{\alpha \in J}$ be a subset of elements of A and let $\{x_\alpha\}_{\alpha \in J}$ be the corresponding set of indeterminates. Then S is a **generating set for** A **as an** R**-algebra** if the evaluation homomorphism

$$\phi_{\{b_\alpha\}} : R[\{x_\alpha\}] \to A$$

is surjective. Equivalently, $S = \{b_\alpha\}$ is a generating set for A if the set of monomials

$$S' = \{b_{\alpha_1}^{e_1} b_{\alpha_2}^{e_2} \cdots b_{\alpha_k}^{e_k}\}$$

where $\alpha_1, \alpha_2, \ldots, \alpha_k$ is a finite set of indices in J and $e_1, e_2, \ldots, e_k \geq 0$ are integers, is a generating set for A as an R-module.

If the generating set S is finite, say, $S = \{b_1, b_2, \ldots, b_n\}$, then A is **finitely generated as an** R**-algebra** and we write $A = R[b_1, b_2, \ldots, b_n]$.

For example, $\{g\}$ is a generating set for the Z-algebra ZC_3, $\langle g \rangle = C_3$, since the evaluation homomorphism $\phi_{\{g\}} : Z[x] \to ZC_3$ is surjective. Every R-algebra A admits a generating set, if necessary one could choose $S = A$.

We aim to characterize every finitely generated R-algebra as a quotient ring.

Proposition 3.29. *Every finitely generated R-algebra A is a quotient of a polynomial ring over R.*

Proof. Let $\{b_1, b_2, \ldots, b_n\}$ be a generating set for A as an R-algebra, let $\{x_1, x_2, \ldots, x_n\}$ be the corresponding set of indeterminates and let $R[x_1, x_2, \ldots, x_n]$ denote the ring of polynomials in the indeterminates x_i. Then the evaluation homomorphism

$$\phi_{\{b_1,\ldots,b_n\}} : R[x_1, \ldots, x_n] \to A$$

is surjective. Let $N = \ker(\phi_{\{b_1,\ldots,b_n\}})$. By Proposition 2.39, $A \cong R[x_1, \ldots, x_n]/N$. □

Let B be an R-algebra and let N be a subset of polynomials in $R[x_1, x_2, \ldots, x_n]$. Then the indexed family $\{b_1, b_2, \ldots, b_n\}$, $b_i \in B$, is a **common zero in B of the polynomials in N** if $f(\{b_1, \ldots, b_n\}) = 0$ for all $f(x_1, \ldots, x_n) \in N$. For example, $\{2, 1/2\}$ is a common zero in $\mathbb{Q}$ of the ideal of polynomials $N = (xy - 1)$ in $\mathbb{Q}[x, y]$. But $\{3\}$ is not a common zero in Z of $\{x + 3, x - 3\} \subseteq Z[x]$.

Proposition 3.30. *Let A be an R-algebra of the form $A = R[x_1, x_2, \ldots, x_n]/N$, where N is an ideal of polynomials and let B be an R-algebra. Then $Hom_{R\text{-}alg}(A, B)$ is in a 1-1 correspondence with the set of common zeros in B of the polynomials in N.*

Proof. Let $\phi : R[x_1, \ldots, x_n]/N \to B$ be an R-algebra homomorphism. Let $\overline{x}_i = x_i \bmod N$ for $1 \le i \le n$. Then ϕ is determined by sending each $\overline{x}_i$ to an element $b_i \in B$. Now for $f(x_1, \ldots, x_n) \in N$,

$$0 = \phi(f(x_1, \ldots, x_n)) = f(\{b_1, \ldots, b_n\}),$$

and so the family $\{b_1, \ldots, b_n\}$ is a common zero in B of the polynomials in N.

Conversely, let $\{b_1, \ldots, b_n\}$ be a common zero of polynomials in N and consider the evaluation homomorphism

$$\phi_{\{b_1,\ldots,b_n\}} : R[x_1, \ldots, x_n] \to B.$$

Now, $\ker(\phi_{\{b_1,\ldots,b_n\}})$ is the ideal of all polynomials for which $\{b_1, \ldots, b_n\}$ is a zero. Thus $\ker(\phi_{\{b_1,\ldots,b_n\}})$ contains N. By the UMPK (Proposition 2.40), there is an R-algebra homomorphism $\phi : A \to B$. □

Let A, B be R-algebras and assume that A is finitely generated. To compute $\text{Hom}_{R\text{-alg}}(A, B)$, we first use Proposition 3.29 to write $A \cong R[x_1, x_2, \ldots, x_n]/N$ for some integer n and some ideal N of $R[x_1, x_2, \ldots, x_n]$. We then apply Proposition 3.30. For example, to compute $\text{Hom}_{Z\text{-alg}}(ZC_3, Z)$, $C_3 = \langle \sigma \rangle$, we note that $ZC_3 \cong Z[x]/(x^3 - 1)$. Now, any Z-algebra homomorphism is of the form $x \mapsto b \in Z$ with $b^3 - 1 = 0$. Thus $\text{Hom}_{Z\text{-alg}}(ZC_3, Z)$ contains exactly one element: $\phi : ZC_3 \to Z$ defined by $\sigma \mapsto 1$.

3.6 Discriminants

If V is a vector space over the field K, then a K-bilinear map $B : V \times V \to K$ of §3.4 is called a bilinear form. This final section of the chapter concerns bilinear forms on a finite dimensional K-vector space V, where K is the field of fractions of an integral domain R. For an R-submodule M of V, which is free over R, and which satisfies $KM = V$, we define the dual module of M^D with respect to a bilinear form B on V. Next we define the discriminant $\text{disc}(M)$ and give some properties of the discriminant. As an example, we let $\mathcal{V}$ be the Klein 4-group, define a bilinear form on $\mathbb{Q}\mathcal{V}$ using the characters of $\mathcal{V}$, and compute $\text{disc}(Z\mathcal{V})$, $Z\mathcal{V}^D$, and $\text{disc}(Z\mathcal{V}^D)$. We shall use discriminants in Chapter 4.

* * *

Let R be an integral domain and let K be its field of fractions (the localization of R at $\{0\}$). Let V be a free K-module of rank k (a vector space V with $\dim(V) = k$).

Definition 3.18. A **bilinear form** B **on** V is a function $B : V \times V \to K$ which satisfies, for $x, y, z \in V$, $a \in K$,

(i) $B(x + y, z) = B(x, z) + B(y, z)$,

(ii) $B(x, y + z) = B(x, y) + B(x, z)$,

(iii) $B(ax, y) = aB(x, y) = B(x, ay)$.

If $B(x, y) = 0$ for all $y \in V$ implies that $x = 0$, and $B(x, y) = 0$ for all $x \in V$ implies that $y = 0$, then B is **non-degenerate**. A bilinear form B is **symmetric** if $B(x, y) = B(y, x)$ for all $x, y \in V$.

Lemma 3.1. *Let B be a non-degenerate bilinear form on V. Let $\{b_1, b_2, \ldots, b_k\}$ be a K-basis for V. Then $\{B(b_1, -), B(b_2, -), \ldots, B(b_k, -)\}$ is a basis for V^*.*

Proof. Clearly, $B(b_i, -) : V \to K$ is a linear transformation, so we only need to check that $\{B(b_i, -)\}_{i=1}^k$ is a linearly independent subset of V^*. Suppose that

$$r_1 B(b_1, -) + r_2 B(b_2, -) + \cdots + r_k B(b_k, -) = 0$$

for $r_1, r_2, \ldots, r_k \in K$. Then for all $y \in V$,

$$\begin{aligned} r_1 B(b_1, y) + r_2 B(b_2, y) + \cdots + r_k B(b_k, y) \\ &= B(r_1 b_1, y) + B(r_2 b_2, y) + \cdots + B(r_k b_k, y) \\ &= B(r_1 b_1 + r_2 b_2 + \cdots + r_k b_k, y) \\ &= 0, \end{aligned}$$

whence $r_1 = r_2 = \cdots = r_k = 0$. $\square$

Let B be a symmetric, non-degenerate bilinear form on V and let M be a free R-submodule of V (viewed as an R-module) which satisfies $KM = V$. Then M has rank k over R. The **dual module of M with respect to $B(x, y)$** is the R-module defined as

$$M^D = \{v \in V : \; B(v, M) \subseteq R\}.$$

The dual module can be identified with the set of R-linear maps $M \to R$, denoted as $\mathrm{Hom}_R(M, R)$, through the isomorphism from M^D to $\mathrm{Hom}_R(M, R)$ given by $\alpha \mapsto B(\alpha, -)$, for $\alpha \in M^D$.

The dual module M^D is free of rank k over R.

Proposition 3.31. *Suppose $\{b_1, b_2, \ldots, b_k\}$ is an R-basis for M. There exists a basis $\{\beta_1, \beta_2, \ldots, \beta_k\}$ for M^D "the dual basis" that satisfies $B(\beta_i, b_j) = \delta_{i,j}$.*

Proof. Since $KM = V$, $\{b_i\}_{i=1}^k$ is also a basis for V. Let $\{f_1, f_2 \ldots, f_k\}$ be the basis for the dual space V^* defined by $f_i(b_j) = \delta_{i,j}$. Let $\{B_{i,j}\}$, $1 \leq i, j \leq k$, be the collection of bilinear forms defined as $B_{i,j}(x, y) = f_i(x) f_j(y)$ for $x, y \in V$. The bilinear form B then can be written uniquely as

$$B = \sum_{i,j=1}^{k} B(b_i, b_j) B_{i,j}.$$

By Lemma 3.1, $\{B(b_i, -)\}_{i=1}^k$ is a basis for V^*. For $1 \leq i \leq k$, one has

$$\begin{aligned} B(b_i, -) &= \sum_{i,j=1}^{k} B(b_i, b_j) B_{i,j}(b_i, -) \\ &= \sum_{j=1}^{k} B(b_i, b_j) f_j, \end{aligned}$$

and so, the $k \times k$ matrix $A = (B(b_i, b_j))$ is the matrix of the change of basis from $\{f_j\}$ to $B(b_i, -)$. Consequently, A is invertible. Let $(\theta_{i,j}) = A^{-1}$ and set

$$\beta_q = \theta_{q,1} b_1 + \theta_{q,2} b_2 + \cdots + \theta_{q,k} b_k,$$

for $1 \leq q \leq k$. Then

$$\begin{aligned} B(\beta_q, b_l) &= \sum_{i,j=1}^{k} B(b_i, b_j) B_{i,j}(\beta_q, b_l) \\ &= \sum_{i=1}^{k} B(b_i, b_l) f_i(\beta_q) \\ &= \sum_{i=1}^{k} B(b_i, b_l) \theta_{q,i} \\ &= \delta_{l,q}. \end{aligned}$$

Thus $\{\beta_1, \beta_2, \ldots, \beta_k\}$ is a basis for M^D. □

From the proof of Proposition 3.31, one has

$$A \begin{pmatrix} \beta_1 \\ \beta_2 \\ \vdots \\ \beta_k \end{pmatrix} = \begin{pmatrix} b_1 \\ b_2 \\ \vdots \\ b_k \end{pmatrix},$$

with $A = (B(b_i, b_j))$. The **discriminant of M over R with respect to B and the basis $\{b_1, b_2, \ldots, b_k\}$** is the R-submodule of K defined as

$$\mathrm{disc}(M) = R\det(A).$$

Now suppose M, N are free R-submodules of V which satisfy $KM = V$, $KN = V$. Then necessarily, both M and N have rank k over R. Let $\{b_i\}_{i=1}^k$ be an R-basis for M and let $\{c_i\}_{i=1}^k$ be an R-basis for N. Then $\{b_i\}$ and $\{c_i\}$ are K-bases for V. Therefore, there exists an invertible $k \times k$ matrix T in $\mathrm{Mat}_k(K)$ for which $T(M) = N$. Let

$$[M : N] = R\det(T).$$

Then $[M : N]$ is an R-submodule of K called the **module index**.

Proposition 3.32. *The module index $[M : N]$ does not depend on the choice of bases for M and N.*

Proof. Let $\{b_i\}$ be an R-basis for M, let $\{c_i\}$ be an R-basis for N and let $T \in \mathrm{Mat}_k(K)$ be so that $T(b_i) = c_i$, for all i. Let $\{b'_i\}$ be some other basis for M and let $\{c'_i\}$ be some other basis for N. Let $S \in \mathrm{Mat}_k(K)$ be so that $S(b'_i) = c'_i$. There exist invertible matrices $X, Y \in \mathrm{Mat}_k(R)$ for which

$X(b_i) = b'_i$, $Y(c_i) = c'_i$ for all i. Now, $Y^{-1}SX(b_i) = c_i$ and so $T = Y^{-1}SX$. One has

$$\det(T) = \det(Y^{-1}SX) = \det(Y^{-1}X)\det(S),$$

with $\det(Y^{-1}X)$ a unit in R. It follows that $R\det(T) = R\det(S)$. □

Proposition 3.33. $[M : N] = [N^D : M^D]$.

Proof. Let $\{b_i\}$ be an R-basis for M and let $\{\beta_i\}$ be the dual basis for M^D. Let $\{c_i\}$ be an R-basis for N and let $\{\gamma_i\}$ be the dual basis for N^D. Let $T \in \mathrm{Mat}_k(K)$, be so that $T(b_i) = c_i$, for all i. Then $T^t(\gamma_i) = \beta_i$, for all i, where T^t is the transpose of T. Thus $[N^D : M^D] = R\det(T^t) = R\det(T) = [M : N]$. □

The module index and the discriminant are related by the following result.

Proposition 3.34. *Suppose M, N are free R-submodules of V with $KN = KM = V$. Then*

$$disc(N) = [M : N]^2 disc(M).$$

Proof. Let $\{b_i\}$ be a basis for M and let $\{c_i\}$ be a basis for N. Suppose T is the matrix for which $T(b_i) = c_i$ for $1 \le i \le k$. Then

$$\begin{aligned}
\mathrm{disc}(N) &= R\det((B(c_i, c_j))) \\
&= R\det((B(T(b_i), T(b_j)))) \\
&= R\det(T \cdot (B(b_i, T(b_j)))) \\
&= R\det(T)\det(T \cdot (B(b_i, b_j))) \\
&= R\det(T)\det(T)\det(B(b_i, b_j)) \\
&= R\det(T)^2\mathrm{disc}(M) \\
&= [M : N]^2\mathrm{disc}(M).
\end{aligned}$$

□

Corollary 3.3. *The discriminant disc(M) does not depend on the choice of basis for M.*

Proof. Indeed, let $\{b_i\}$ and $\{c_i\}$ be two bases for the R-module M. Let $T \in \mathrm{Mat}_k(K)$ be so that $T(b_i) = c_i$ for all i. Then $T \in \mathrm{Mat}_k(R)$ with $\det(T) \in U(R)$. Consequently, $[M : M] = R$. Now by Proposition 3.34, disc(M) computed using $\{b_i\}$ is the same as disc(M) computed using $\{c_i\}$. □

The next proposition provides an efficient way to show that two modules are equal.

Proposition 3.35. *Let M, N be free R-modules with $KM = KN = V$. Suppose $N \subseteq M$ with disc$(M) =$ disc(N). Then $M = N$.*

Proof. First note that $N \subseteq M$ implies that $[M : N]$ is an ideal of R. By Proposition 3.34, $[M : N]^2 = R$. Hence the module index $[M : N] = R$ and the matrix T, $T(M) = N$, has inverse T^{-1} in $\mathrm{Mat}_k(R)$. Thus $M = T^{-1}(N) \subseteq N$ and hence, $N = M$. □

We close this section with some computations. Let $R = Z$, $K = \mathbb{Q}$, and let $\mathcal{V} = \{e, a, b, c\}$ denote the Klein 4-group. Let $V = \mathbb{Q}\mathcal{V}$. To simplify notation, we rename the elements of $\mathcal{V}$ as $g_1 = e$, $g_2 = a$, $g_3 = b$, $g_4 = c$. From §2.3, the characters of $\mathcal{V}$ are ν_1, ν_2, ν_3, ν_4.

The trace map $\mathrm{tr} : \mathbb{Q}\mathcal{V} \to \mathbb{Q}$ is defined as $\mathrm{tr}(x) = \sum_{i=1}^{4} \nu_i(x)$ for $x \in \mathbb{Q}\mathcal{V}$. Note that $\mathrm{tr}(\sum_{i=1}^{4} a_i g_i) = 4a_1$ for $a_i \in \mathbb{Q}$. Let

$$B(x, y) = \mathrm{tr}(xy).$$

Then B is a symmetric, non-degenerate bilinear form on $\mathbb{Q}\mathcal{V}$ (see §3.7, Exercise 23). We compute disc$(Z\mathcal{V})$, $Z\mathcal{V}^D$, and disc$(Z\mathcal{V}^D)$ with respect to B.

Since $\{g_1, g_2, g_3, g_4\}$ is a Z-basis for $Z\mathcal{V}$, one easily computes disc$(Z\mathcal{V}) = 256Z$. For the other calculations, we first consider the group ring $\mathbb{Q}\mathcal{V}$. The set $\{g_1, g_2, g_3, g_4\}$ is a $\mathbb{Q}$-basis for $\mathbb{Q}\mathcal{V}$, and with respect to this basis the vectors are multiplied according to the group product on $\mathcal{V}$. But is there a basis E for $\mathbb{Q}\mathcal{V}$ for which the coordinate vectors v_E are multiplied component-wise? In other words, we seek a basis E so that

$$v_E w_E = \left(\sum_{i=1}^{4} v_i \alpha_i\right)\left(\sum_{j=1}^{4} w_j \alpha_j\right) = \sum_{i=1}^{4} v_i w_i \alpha_i.$$

Such a basis does exist and has the form $E = \{\alpha_1, \alpha_2, \alpha_3, \alpha_4\}$ where

$$\alpha_i = \frac{1}{4}\sum_{m=1}^{4} \nu_i(g_m) g_m,$$

for $i = 1, 2, 3, 4$. These elements are the minimal idempotents of $\mathbb{Q}\mathcal{V}$, see [Serre (1977), §2.3]. The existence of this basis implies that as rings,

$$\mathbb{Q}\mathcal{V} = \bigoplus_{i=1}^{4} \mathbb{Q}\alpha_i \cong \mathbb{Q} \times \mathbb{Q} \times \mathbb{Q} \times \mathbb{Q}.$$

We now proceed with the computation of $Z\mathcal{V}^D$. By definition,

$$Z\mathcal{V}^D = \{u \in \mathbb{Q}\mathcal{V} : \operatorname{tr}(uZ\mathcal{V}) \subseteq Z\}.$$

We have

$$\operatorname{tr}\left(\frac{g_m}{4} g_n\right) = \delta_{m,n}$$

for $m, n = 1, 2, 3, 4$, and so

$$S = \left\{\frac{g_1}{4}, \frac{g_2}{4}, \frac{g_3}{4}, \frac{g_4}{4}\right\}$$

is the dual basis; $Z\mathcal{V}^D$ is the free rank 4 Z-module on the basis S.

In order to compute $\operatorname{disc}(Z\mathcal{V}^D)$ we need to view $Z\mathcal{V}^D$ as an Z-subalgebra of $\mathbb{Q}\mathcal{V}$. On $Z\mathcal{V}^D$ we define addition to be the addition on $Z\mathcal{V}^D$ as an R-submodule of $\mathbb{Q}\mathcal{V}$, and we define multiplication $*$ on $Z\mathcal{V}^D$ as follows. For $g_n \in \mathcal{V}$, $1 \leq l, m \leq 4$,

$$\begin{aligned}\left(\frac{g_l}{4} * \frac{g_m}{4}\right)(g_n) &= \frac{g_l}{4}(g_n)\frac{g_m}{4}(g_n) \\ &= \delta_{l,n}\delta_{m,n} \\ &= \delta_{l,m}\delta_{m,n} \\ &= \delta_{l,m}\frac{g_m}{4}(g_n).\end{aligned}$$

Thus, S is the collection of minimal idempotents of the Z-algebra $Z\mathcal{V}^D$ endowed with the binary operations $+$ and $*$.

Proposition 3.36. *$Z\mathcal{V}^D \cong \bigoplus_{i=1}^{4} Z\alpha_i$, as Z-algebras.*

Proof. The map

$$\phi : Z\mathcal{V}^D \to \bigoplus_{i=1}^{4} Z\alpha_i,$$

defined by $\frac{g_i}{4} \mapsto \alpha_i$ is an Z-algebra isomorphism. □

We identify $Z\mathcal{V}^D$ with $\bigoplus_{i=1}^{4} Z\alpha_i$ through the isomorphism of Proposition 3.36. Thus we consider $\{\alpha_1, \alpha_2, \alpha_3, \alpha_4\}$ as a Z-basis for $Z\mathcal{V}^D \subseteq \mathbb{Q}\mathcal{V}$. One now easily computes $\operatorname{tr}(\alpha_i\alpha_j) = \delta_{i,j}$ and thus, $\operatorname{disc}(Z\mathcal{V}^D) = Z$.

3.7 Exercises

Exercises for §3.1

(1) Let V be a finite dimensional vector space over a field K with $\dim(V) = n$. Let $\{v_1, v_2, \ldots, v_m\}$ be a subset of V with $m > n$. Show that S is linearly dependent.
(2) Prove Proposition 3.6.
(3) Let V be a finite dimensional vector space over the field K with linear dual V^*. The **double dual** is the dual space $V^{**} = (V^*)^*$. Show that there is a "natural" isomorphism $\phi : V \to V^{**}$ defined as $\phi(a)(f) = f(a)$, for $a \in V$, $f \in V^*$.

Exercises for §3.2

(4) Let $\phi : M \to M'$ be an isomorphism of free R-modules. Suppose that $\{m_1, m_2, \ldots, m_k\}$ is a basis for M. Show that $\{\phi(m_1), \phi(m_2), \ldots, \phi(m_k)\}$ is a basis for M'.
(5) Let R be a ring and let M be an R-module. Prove that scalar multiplication $R \times M \to M$, $(r, m) \mapsto rm$, defines an R-module homomorphism $\phi : R \to M$, $r \mapsto rm$.
(6) Let R be an integral domain and let I be an ideal of R which is free of finite rank as an R-module. Show that I is a principal ideal of R.
(7) Show that an $\mathbb{R}$-submodule of $\mathbb{R}^n$ is isomorphic to $\mathbb{R}^m$ for some $m \leq n$.
(8) Prove that Z_n is finitely generated as a Z-module but not free over Z.
(9) Let R be an integral domain and let M be a free R-module. Suppose that $rm = 0$ for some $r \in R$, $m \in M$. Prove that either $r = 0_R$ or $m = 0$.
(10) A group is **simple** if it is non-trivial and has no proper non-trivial normal subgroups. A non-zero R-module M is **simple** if it has no proper non-zero submodules. Let G be an abelian group which is simple as a Z-module. Show that G is a simple group.

Exercises for §3.3

(11) Give an example of a short exact sequence of R-modules that is not split.
(12) Show that Z_n is not projective as a Z-module.
(13) Prove the converse of Proposition 3.16.
(14) Prove Corollary 3.2.

Exercises for §3.4

(15) Let M, N be R-modules and let $\tau : M \otimes_R N \to N \otimes_R M$ be the **twist map** defined as $\tau(a \otimes b) = b \otimes a$ for $a \in M$, $b \in N$. Prove that τ is an isomorphism of R-modules.

(16) Let M_1, M_2, N be R-modules. Prove that $(M_1 \oplus M_2) \otimes N \cong (M_1 \oplus N) \otimes (M_2 \oplus N)$.

(17) Prove that $Z_m \otimes_Z Z_n = \{0\}$ if and only if $\gcd(m, n) = 1$.

(18) Find an element of $Z[x] \otimes_Z Z[x]$ that can not be written in the form $f(x) \otimes g(x)$ for $f(x), g(x) \in Z[x]$.

(19) Let R be an integral domain let P, Q be prime ideals of R and suppose that M is a finitely generated projective R-module.

(a) Prove that M_P and M_Q are free modules over R_P and R_Q, respectively.

(b) Show that $\operatorname{rank}(M_P) = \operatorname{rank}(M_Q)$.

Exercises for §3.5

(20) Show that $R[x] \otimes_R R[x] \cong R[x, y]$ as R-algebras.

(21) Compute $\operatorname{Hom}_{\mathbb{Q}\text{-alg}}(\mathbb{Q}[x], \mathbb{R})$.

(22) Let $p > 2$ be a prime number.

(a) Compute $\operatorname{Hom}_{\mathbb{R}\text{-alg}}(\mathbb{R}[x]/(x^p - 1), \mathbb{R})$.

(b) Compute $\operatorname{Hom}_{\mathbb{C}\text{-alg}}(\mathbb{C}[x]/(x^p - 1), \mathbb{R})$.

Exercises for §3.6

(23) As in §3.6, let $\operatorname{tr} : \mathbb{Q}\mathcal{V} \to \mathbb{Q}$ be the trace map defined as $\operatorname{tr}(x) = \sum_{i=1}^{4} \nu^i(x)$ for $x \in \mathbb{Q}\mathcal{V}$. Let $B(x, y) = \operatorname{tr}(xy)$. Show that B is a symmetric, non-degenerate bilinear form on $\mathbb{Q}\mathcal{V}$.

(24) Let R be an integral domain with field of fractions K. Let V be a vector space over K and let M be a free R-module with $KM = V$. Prove that $(M^D)^D = M$.

(25) Let R be an integral domain with field of fractions K. Let V be a vector space over K, and let M, N be free R-modules with $KM = V$ and $KN = V$. Prove that $N \subseteq M$ if and only if $M^D \subseteq N^D$.

(26) Let R be an integral domain with field of fractions K. Let V be a vector space over K, and let M, N be free R-modules with $KM = V$ and $KN = V$ and $N \subseteq M$. Show that $\operatorname{disc}(N) \subseteq \operatorname{disc}(M)$.

(27) Let R be an integral domain with field of fractions K. Let V be a vector space over K and let M be a free R-module with $KM = V$.

Suppose that $V = V_1 \oplus V_2$ and $M = M_1 \oplus M_2$ with $KM_1 = V_1$ and $KM_2 = V_2$. Prove that $\mathrm{disc}(M) = \mathrm{disc}(M_1)\mathrm{disc}(M_2)$.

Questions for Further Study

(1) Let p be a prime number, let ζ_p denote a primitive pth root of unity, and let $K = \mathbb{Q}(\zeta_p)$ denote the simple algebraic field extension, cf. §4.1. Then the ring of integers of K is the Z-algebra $R = Z[\zeta_p]$, see Proposition 4.22; R is an integral domain with field of fractions K.

(a) Compute the minimal idempotents for the group ring KC_p, where C_p denotes the cyclic group of order p.

(b) Compute $\mathrm{disc}(RC_p)$ and $\mathrm{disc}(RC_p^D)$.

Chapter 4

Simple Algebraic Extension Fields

This chapter concerns simple algebraic extensions of a field F. We specialize to the case where $F = K$ is an intermediate field $\mathbb{Q} \subseteq K \subseteq \mathbb{C}$. We construct the Galois group G of the splitting field L/K of an irreducible monic (hence, separable) polynomial $p(x) \in K[x]$. We give some examples of Galois groups and state the Fundamental Theorem of Galois Theory which says that there is a lattice-inverting bijection between the set of intermediate fields between K and L and subgroups of G. A consequence of the Fundamental Theorem is that since G is finite, there is only a finite number of intermediate fields L', $K \subseteq L' \subseteq L$. This is remarkable considering that K is an infinite field.

The next part of the chapter (§4.3–§4.6) constitutes an introduction to algebraic number theory. We define the ring of integers R of a simple algebraic extension K of $\mathbb{Q}$, and show that every non-zero ideal of R contains a $\mathbb{Q}$-basis for K, specifically, R is a free Z-module of rank $n = [K : \mathbb{Q}]$. We show how to construct the ring of integers of $\mathbb{Q}(\sqrt{m})$ and compute the ring of integers of $\mathbb{Q}(\zeta_p)$, where ζ_p is a primitive pth root of unity. We employ Dirichlet's Unit Theorem to construct the group of units of the ring of integers R. Specifically, we compute $U(R)$ where R is the ring of integers of a quadratic field extension and where R is the ring of integers of $\mathbb{Q}(\zeta_p)$.

We then show that the ring of integers R is a Noetherian ring by showing that R has the ACC. Moreover, R has additional properties: it is a Dedekind domain. Arguably, the most important property of a Dedekind domain is that its localization at any prime ideal is a PID. We use this fact to construct the class group of a Dedekind domain R, which measures the extent to which R fails to be a PID. We show that every non-zero proper ideal of R factors uniquely into a product of prime ideals, thus generalizing the familiar notion that each integer $m \in Z$ factors uniquely into a product primes $m = p_1^{e_1} p_2^{e_2} \cdots p_k^{e_k}$.

The final section of the chapter concerns field extensions of $\mathbb{Q}_p$, the field of p-adic rationals that was constructed in §2.7. We construct finite field extensions of $\mathbb{Q}_p$ by extending the notion of p-adic absolute value and using the method of §2.7. Of course, another way to construct an extension field of $\mathbb{Q}_p$ (or any field for that matter) is to take an irreducible polynomial $p(x) \in \mathbb{Q}_p[x]$ of degree n, and "invent" a root $\alpha = x + (p(x))$ of $p(x)$ in the field $\mathbb{Q}_p[x]/(p(x))$. Then $\mathbb{Q}_p[x]/(p(x)) \cong \mathbb{Q}_p(\alpha)$ will be a simple algebraic extension of $\mathbb{Q}_p$, a finite dimensional vector space over $\mathbb{Q}_p$ on the basis $\{1, \alpha, \alpha^2, \ldots, \alpha^{n-1}\}$. This idea of inventing roots will be used in Chapter 5 (Finite Fields).

4.1 Simple Algebraic Extensions

In this section we define the notion of a simple algebraic extension of a field F. We specialize to the case where $F = K$ is an intermediate field $\mathbb{Q} \subseteq K \subseteq \mathbb{C}$. Then the Fundamental Theorem of Algebra applies to show that all simple algebraic extensions of K are subfields of $\mathbb{C}$. Moreover, L is a simple algebraic extension of K if and only if L is a finite dimensional vector space over K. We show that a simple algebraic extension L of K of degree n determines a set of n distinct embeddings $\tau_i : L \to \mathbb{C}$, $1 \le i \le n$. Furthermore, if E is a simple algebraic extension of L of degree m, then each embedding $\tau_i : L \to \mathbb{C}$ extends to m distinct embeddings $\tau_{i,j} : E \to \mathbb{C}$, $1 \le j \le m$. From this extension theorem, we define the norm and trace maps for extensions L/K.

* * *

Let E/F be a field extension with $\alpha \in E$. Assume that α is a zero of a non-constant polynomial $f(x) \in F[x]$. Let $\phi_\alpha : F[x] \to E$ be the evaluation homomorphism.

Proposition 4.1. *Let $M = ker(\phi_\alpha)$. Then $F[x]/M$ is a field and M is a non-zero maximal ideal of $F[x]$.*

Proof. By Proposition 2.39, $F[x]/M \cong \phi_\alpha(F[x]) \subseteq E$, and so, $F[x]/M$ is isomorphic to a subring of a field, and hence must be an integral domain. Thus M is prime by Proposition 2.35, and non-trivial since $f(x) \in M$. Since $F[x]$ is a PID (Proposition 2.20), $M = (p(x))$ for some monic polynomial

$$p(x) = a_0 + a_1 x + \cdots + a_{n-1} x^{n-1} + x^n$$

of degree n. It follows by Proposition 2.25 that $p(x)$ is irreducible, and thus $M = (p(x))$ is maximal. Consequently, $F[x]/M$ is a field. $\square$

We define the polynomial $p(x)$ of Proposition 4.1 to be the **irreducible polynomial for α over F**, which we denote as $\text{irr}(\alpha, F)$.

But what is the structure of the field $F[x]/(p(x))$?

Proposition 4.2. *The field $F[x]/(p(x)) \cong \phi_\alpha(F[x])$ is a finite dimensional vector space over F of dimension $n = \deg(p(x))$ on the "power basis" $\{1, \alpha, \alpha^2, \ldots, \alpha^{n-1}\}$.*

Proof. First note that $F[x]$ is a vector space over F, $(p(x))$ is a subspace and the quotient space $F[x]/(p(x))$ is a vector space over F. Moreover, the map ϕ_α is a linear transformation of F-vector spaces, and consequently, $F[x]/(p(x)) \cong \phi_\alpha(F[x])$ is an isomorphism of F-vector spaces. Thus, it is natural to characterize the field $F[x]/(p(x))$ as the collection of all polynomials with coefficients in F evaluated at α, where $p(\alpha) = 0$, or equivalently, where the powers of α satisfy the relation

$$\alpha^n = -a_{n-1}\alpha^{n-1} - \cdots - a_2\alpha^2 - a_1\alpha - a_0.$$

Thus any power α^m with $m \geq n$ can be written in terms of lower powers $\{1, \alpha, \alpha^2, \ldots, \alpha^{n-1}\}$. Consequently, every element of $\phi_\alpha(F[x])$ has a representation as a linear combination

$$r_0 + r_1\alpha + r_2\alpha^2 + \cdots + r_{n-1}\alpha^{n-1}, \quad r_i \in F,$$

that is, $\{1, \alpha, \alpha^2, \ldots, \alpha^{n-1}\}$ is a generating set for $\phi_\alpha(F[x])$ over F. Moreover, $\{1, \alpha, \alpha^2 \ldots, \alpha^{n-1}\}$ is a basis for $\phi_\alpha(F[x])$ over F since $p(x)$ is irreducible. We conclude that $F[x]/(p(x)) \cong \phi_\alpha(F[x])$ is an n-dimensional vector space over the field F with the power basis $\{1, \alpha, \alpha^2, \ldots, \alpha^{n-1}\}$. $\square$

To illustrate Proposition 4.2, let $F = \mathbb{Q}$, $E = \mathbb{R}$ with $\alpha = \sqrt[3]{2}$ (the real zero of $x^3 - 2 \in \mathbb{Q}[x]$.) Then there is an evaluation homomorphism $\phi_{\sqrt[3]{2}} : \mathbb{Q}[x] \to \mathbb{R}$ with $\ker(\phi_{\sqrt[3]{2}}) = (x^3 - 2)$. We have

$$\mathbb{Q}[x]/(x^3 - 2) \cong \phi_{\sqrt[3]{2}}(\mathbb{Q}[x])$$

and $\phi_{\sqrt[3]{2}}(\mathbb{Q}[x])$ consists of all quantities of the form

$$a_0 + a_1(\sqrt[3]{2}) + a_2(\sqrt[3]{4}), \; a_0, a_1, a_2 \in \mathbb{Q}.$$

As a vector space over $\mathbb{Q}$, $\phi_{\sqrt[3]{2}}(\mathbb{Q}[x])$ has dimension 3 with basis $\{1, \sqrt[3]{2}, \sqrt[3]{4}\}$.

As a field, every non-zero element in $\phi_{\sqrt[3]{2}}(\mathbb{Q}[x])$ has a multiplicative inverse. How does one compute $(2+\sqrt[3]{2})^{-1}$? Consider the polynomial $2+x$. Now a greatest common divisor of $2+x$ and x^3-2 is 1, and so, by Proposition 2.21, there exists polynomials $r(x)$ and $s(x)$ in $\mathbb{Q}[x]$ for which

$$(2+x)r(x)+(x^3-2)s(x)=1.$$

Thus $(2+\sqrt[3]{2})r(\sqrt[3]{2})=1$, so that $r(\sqrt[3]{2})$ is the inverse of $2+\sqrt[3]{2}$. (In fact, $r(\sqrt[3]{2})=(4-2\sqrt[3]{2}+\sqrt[3]{4})/10$.)

For another example we take $F=\mathbb{Q}$, $E=\mathbb{C}$ with $\alpha=\zeta_p$, a primitive pth root of unity. The evaluation homomorphism $\phi_{\zeta_p}:\mathbb{Q}[x]\to\mathbb{C}$ has kernel $\ker(\phi_{\zeta_p})=(p(x))$ with $p(x)=x^{p-1}+x^{p-2}+\cdots+x^2+x+1$. Note that $p(x)$ is irreducible by Eisenstein's criterion (Proposition 2.8). We have $\mathbb{Q}[x]/(p(x))\cong\phi_{\zeta_p}(\mathbb{Q}[x])$, and $\phi_{\zeta_p}(\mathbb{Q}[x])$ consists of all vectors of the form

$$a_0+a_1\zeta_p+a_2\zeta_p^2+\cdots+a_{p-2}\zeta_p^{p-2}$$

for $a_i\in\mathbb{Q}$. As a vector space over $\mathbb{Q}$, $\phi_{\zeta_p}(\mathbb{Q}[x])$ has dimension $p-1$ with basis $\{1,\zeta_p,\zeta_p^2,\dots,\zeta_p^{p-2}\}$.

Definition 4.1. The field $\phi_\alpha(F[x])$ of Proposition 4.1 is the **simple algebraic extension of** F **by** α and is denoted as $F(\alpha)$. The **degree** of $F(\alpha)$ over K, denoted by $[F(\alpha):F]$, is the dimension of the vector space $F(\alpha)$ over F. Thus $[F(\alpha):F]$ is the degree of the irreducible polynomial $\mathrm{irr}(\alpha,F)$ of α over F.

Proposition 4.3. *Let $F(\alpha)$ be a simple algebraic extension of F with $[F(\alpha):F]=n$. Then every element β in $F(\alpha)$ is a root of a monic polynomial of degree $\le n$ with coefficients in F.*

Proof. Since $[F(\alpha):F]=n$, there is a smallest integer $m\le n$ for which the subset $\{1,\beta,\beta^2,\dots,\beta^m\}$ is linearly dependent (cf. Proposition 3.5). Thus, there exist elements $a_0,a_1,\dots,a_m$ of F for which

$$\beta^m=a_0+a_1\beta+\cdots+a_{m-1}\beta^{m-1}.$$

And so, β is a zero of the polynomial $f(x)=-a_0-a_1x-\cdots-a_{n-1}x^{m-1}+x^m$ with $m\le n$. □

In fact, $\beta\in F(\alpha)$ is a root of an irreducible monic polynomial over F.

Proposition 4.4. *Let $F(\alpha)$ be a simple algebraic extension of F. Let $\beta\in F(\alpha)$. Then there exists an irreducible monic polynomial $p(x)\in F[x]$ for which $p(\beta)=0$.*

Proof. Indeed, let $\phi_\beta : F[x] \to F(\alpha)$ be the evaluation homomorphism, and take $p(x)$ to be the generator of the principal ideal $\ker(\phi_\beta)$. □

Proposition 4.5. *Let $K = F(\alpha)$ be a simple algebraic extension of F of degree $[K : F]$, and let $L = K(\beta)$ be a simple algebraic extension of K of degree $[L : K]$. Then $L = K(\beta) = F(\alpha)(\beta)$ is a finite extension of F of degree $[L : F] = [L : K][K : F]$. We write $L = F(\alpha, \beta)$.*

Proof. Let $m = [K : F]$, $n = [L : K]$. Then $\{1, \alpha, \alpha^2, \dots, \alpha^{m-1}\}$ is a F-basis for K and that $\{1, \beta, \beta^2, \dots, \beta^{n-1}\}$ is an K-basis for L. Clearly, $\{\alpha^i\beta^j\}$, $0 \le i \le m-1$, $0 \le j \le n-1$, is a generating set for L over F, so it remains to show that $\{\alpha^i\beta^j\}$ is linearly independent. Suppose that $\sum_{i=0}^{m-1}\sum_{j=0}^{n-1} r_{i,j}\alpha^i\beta^j = 0$ for $r_{i,j} \in F$. Then $\sum_{j=0}^{n-1}\left(\sum_{i=0}^{m-1} r_{i,j}\alpha^i\right)\beta^j = 0$, and since $\{1, \beta, \beta^2, \dots, \beta^{n-1}\}$ is an $F(\alpha)$-basis for L, $\sum_{i=1}^{m-1} r_{i,j}\alpha^i = 0$, for $0 \le j \le n-1$. It follows that $r_{i,j} = 0$ for all i, j. The result follows. □

For the remainder this section we assume that K is an intermediate field $\mathbb{Q} \subseteq K \subseteq \mathbb{C}$, thus any polynomial $f(x) \in K[x]$ has coefficients in $\mathbb{C}$.

Proposition 4.6 (Fundamental Theorem of Algebra). *Let $f(x)$ be a non-constant polynomial in $\mathbb{C}[x]$. Then there is a zero of $f(x)$ in $\mathbb{C}$.*

Proof. We choose to omit a proof here, but see [Rotman (2002), Theorem 4.49]. □

Proposition 4.7. *Let $f(x)$ be a non-constant monic polynomial of degree n in $\mathbb{C}[x]$. Then*

$$f(x) = (x - \alpha_1)(x - \alpha_2)\cdots(x - \alpha_n),$$

where $\alpha_1, \alpha_2, \dots, \dots, \alpha_n$ are complex numbers.

Proof. By Proposition 4.6, there exists a root α_1 of $f(x)$ in $\mathbb{C}$. By the Factor Theorem

$$f(x) = (x - \alpha_1)q_1(x),$$

for $q_1(x) \in \mathbb{C}[x]$. Since $f(x)$ is monic, either $q_1(x) = 1$ or $q_1(x)$ is non-constant. Assume that $q_1(x)$ is non-constant. Then by Proposition 4.6 and the Factor Theorem, there exists $\alpha_2 \in \mathbb{C}$ with

$$f(x) = (x - \alpha_1)(x - \alpha_2)q_2(x)$$

where either $q_2(x) = 1$ or $q_2(x)$ is non-constant. Continuing in this manner one obtains

$$f(x) = (x - \alpha_1)(x - \alpha_2) \cdots (x - \alpha_n),$$

as required. □

Proposition 4.8. *Let $p(x)$ be an irreducible monic polynomial of degree $n \geq 1$ in $K[x]$. Then the zeros of $p(x)$ are distinct.*

Proof. By Proposition 4.7, all of the zeros of $p(x)$ are complex numbers. Let α be any one of these zeros and let $\phi_\alpha : K[x] \to \mathbb{C}$ be the evaluation homomorphism defined as $f(x) \mapsto f(\alpha)$. Then $\ker(\phi_\alpha) = (p(x))$, and so, $p(x)$ is the monic polynomial of smallest degree for which α is a zero.

By way of contradiction, assume that α has multiplicity ≥ 2. Since K has characteristic 0, the formal derivative $p'(x)$ is a non-constant polynomial of degree $n-1 < n$ for which α is a root. This contradicts that $p(x)$ is a polynomial of smallest degree for which $p(\alpha) = 0$, thus the zeros of $p(x)$ are distinct. □

Now, let $\alpha, \beta \in \mathbb{C}$, let $L = K(\alpha)$ be the simple algebraic extension of K and let $E = L(\beta) = K(\alpha)(\beta) = K(\alpha, \beta)$ be the simple algebraic extension of L. By Proposition 4.5, E is a finite extension of K of degree $[E : L][L : K]$. Remarkably, E is a simple algebraic extension of K.

Proposition 4.9. *Let $E = K(\alpha, \beta)$ as above. Then there exists an element $\eta \in E$ for which $E = K(\eta)$.*

Proof. Let $q(x) = \mathrm{irr}(\alpha, K)$ be the irreducible polynomial of α. The roots of $q(x)$ are distinct (Proposition 4.8), and we may list them as $\alpha = \alpha_1, \alpha_2, \ldots, \alpha_l$. Let $r(x) = \mathrm{irr}(\beta, K)$ be the irreducible polynomial of β. Again the roots of $r(x)$ are distinct, and we write $\beta = \beta_1, \beta_2, \ldots, \beta_k$.

Since K has characteristic 0, there exists an element $t \in K$ for which $t \neq (\alpha_i - \alpha)/(\beta - \beta_j)$ for all i, j, $1 \leq i \leq l$, $2 \leq j \leq k$. Set $\eta = \alpha + t\beta$. Clearly, $K(\eta) \subseteq K(\alpha, \beta)$, so it remains to show the reverse inclusion. To this end, observe that

$$\eta - t\beta_j \neq \alpha_i,$$

for all $1 \leq i \leq l$, $2 \leq j \leq k$. Note that $q(\eta - tx) \in K(\eta)[x]$ and put $f(x) = q(\eta - tx)$. Then $f(\beta) = q(\eta - t\beta) = q(\alpha) = 0$, and for $2 \leq j \leq k$, $f(\beta_j) = q(\eta - t\beta_j) \neq 0$ since $\eta - t\beta_j \neq \alpha_i$ for all $1 \leq i \leq l, 2 \leq j \leq k$, and the α_i are the only zeros of $q(x)$. It follows that $f(x)$ and $r(x)$ share a common factor in $K(\eta)[x]$ which must be $x - \beta$. Thus $\beta \in K(\eta)$. Consequently, $\alpha = \eta - t\beta \in K(\eta)$, and so $K(\alpha, \beta) \subseteq K(\eta)$. □

For a field K with $\mathbb{Q} \subseteq K \subseteq \mathbb{C}$, the notions of simple algebraic extension and finite extension of fields are equivalent.

Proposition 4.10. *L/K is a simple algebraic extension if and only if L/K is a finite extension of fields.*

Proof. It is immediate that the simple algebraic extension $L = K(\alpha)$ is a finite extension of fields. For the converse suppose that L/K is a finite extension of fields. Let $\{b_1, b_2, \ldots, b_n\}$ be a K-basis for L. Then $L = K(b_1)(b_2)\cdots(b_n)$ and by repeated uses of Proposition 4.9 one arrives at $L = K(\alpha)$ for some $\alpha \in L$. □

Let K be a field with $\mathbb{Q} \subseteq K \subseteq \mathbb{C}$, and let $L = K(\alpha)$, $\alpha \in \mathbb{C}$, be a simple algebraic extension of K with $q(x) = \mathrm{irr}(\alpha, K)$ and $n = \deg(q(x)) = [L : K]$. Then by Proposition 4.7 and Proposition 4.8, the roots of $q(x)$ are distinct complex numbers which can be listed as $\alpha = \alpha_1, \alpha_2, \ldots, \alpha_n$. For each i, $1 \le i \le n$, there exists an injective homomorphism of rings

$$g_i : L \to \mathbb{C}$$

defined as $g_i(\alpha) = \alpha_i$. Note that g_i fixes K (cf. §2.8, Exercise 68). These homomorphisms are the **embeddings of L into the complex numbers**.

Assume $K \subseteq \mathbb{R}$. If $\alpha_i \in \mathbb{R}$, then g_i is a **real embedding**, if $\alpha_i \in \mathbb{C}\backslash\mathbb{R}$, then g_i is a **complex embedding**. Moreover, if $K \subseteq \mathbb{R}$ and $\alpha_i \in \mathbb{C}\backslash\mathbb{R}$ for some $1 \le i \le n$, then there exists an integer $j \ne i$ so that the complex conjugate $\tilde{\alpha}_i = \alpha_j$. In other words, zeros in $\mathbb{C}\backslash\mathbb{R}$ occur in conjugate pairs. Thus for $K \subseteq \mathbb{R}$, if r denotes the number of real embeddings and s denotes the number of pairs of conjugate complex embeddings, then $n = r + 2s$.

Now, let K be a field with $\mathbb{Q} \subseteq K \subseteq \mathbb{C}$, and let $\beta \in L = K(\alpha)$, $\alpha \in \mathbb{C}$. By Proposition 4.4 there exists a monic irreducible polynomial $p(x) \in K[x]$ for which $p(\beta) = 0$; $K(\beta)$ is a finite extension of K of degree $s = \deg(p(x)) = [K(\beta) : K]$; $K(\alpha) = K(\beta)(\alpha)$ is a finite extension of $K(\beta)$ of degree $t = [K(\alpha) : K(\beta)]$.

Let $\beta = \beta_1, \beta_2, \ldots, \beta_s$ denote the distinct roots of $p(x)$ corresponding to embeddings

$$\tau_i : K(\beta) \to \mathbb{C}, \quad \beta \mapsto \beta_i,$$

$1 \le i \le s$.

Proposition 4.11. *Let $\tau_i : K(\beta) \to \mathbb{C}$, $1 \le i \le s$, denote the set of s embeddings as above. Then each τ_i extends to a set of t embeddings $\tau_{i,j} : K(\alpha) \to \mathbb{C}$, $1 \le j \le t$.*

Proof. Let

$$q(x) = a_0 + a_1x + a_2x^2 + \cdots + a_{t-1}x^{t-1} + x^t, \quad a_i \in K(\beta),$$

be the irreducible monic polynomial of α over $K(\beta)$ of degree $t = [K(\alpha) : K(\beta)]$. For each i, $1 \le i \le s$, let

$$q^{\tau_i}(x) = \tau_i(a_0)+\tau_i(a_1)x+\tau_i(a_2)x^2+\cdots+\tau_i(a_{t-1})x^{t-1}+x^t, \quad \tau_i(a_j) \in K(\beta_i),$$

and let $\gamma_{i,1}, \gamma_{i,2}, \ldots, \gamma_{i,t}$ denote the distinct roots of $q^{\tau_i}(x)$ in $\mathbb{C}$.

Now every element of $K(\alpha)/K(\beta)$ can be written in the form $f(\alpha)$ where

$$f(x) = b_0 + b_1x + \cdots + b_rx^r,$$

$r \ge 0$, is a polynomial in $K(\beta)[x]$. So for each pair i, j, $1 \le i \le s$, $1 \le j \le t$, there is a map

$$\tau_{i,j} : K(\alpha) \to \mathbb{C},$$

defined by

$$b_0 + b_1\alpha + \cdots + b_r\alpha^r \mapsto \tau_i(b_0) + \tau_i(b_1)\gamma_{i,j} + \cdots + \tau_i(b_r)\gamma_{i,j}^r.$$

As one can check these are embeddings which extend τ_i. □

To illustrate Proposition 4.11, we consider the finite extension of fields $\mathbb{Q}(\sqrt{2}, \sqrt{3})/\mathbb{Q}$ of degree 4 over $\mathbb{Q}$. An application of Proposition 4.9 shows that $\mathbb{Q}(\sqrt{2}, \sqrt{3}) = \mathbb{Q}(\sqrt{2} + \sqrt{3})$. Let $\beta = \sqrt{2}$ and consider the tower of fields $\mathbb{Q} \subseteq \mathbb{Q}(\sqrt{2}) \subseteq \mathbb{Q}(\sqrt{2}+\sqrt{3})$. There are two embeddings of $\mathbb{Q}(\sqrt{2})$ into $\mathbb{C}$:

$$\tau_1 : \mathbb{Q}(\sqrt{2}) \to \mathbb{C}, \quad \sqrt{2} \mapsto \sqrt{2},$$

$$\tau_2 : \mathbb{Q}(\sqrt{2}) \to \mathbb{C}, \quad \sqrt{2} \mapsto -\sqrt{2}.$$

Note that the irreducible polynomial of $\sqrt{2} + \sqrt{3}$ over $\mathbb{Q}(\sqrt{2})$ is

$$q(x) = x^2 - 2\sqrt{2}x - 1.$$

One has $q^{\tau_1}(x) = q(x)$ and the roots of $q(x)$ are $\gamma_{1,1} = \sqrt{2} + \sqrt{3}$ and $\gamma_{1,2} = \sqrt{2} - \sqrt{3}$. Thus τ_1 extends to two embeddings $\mathbb{Q}(\sqrt{2} + \sqrt{3}) \to \mathbb{C}$ defined by

$$\tau_{1,1}(\sqrt{2} + \sqrt{3}) = \sqrt{2} + \sqrt{3},$$

$$\tau_{1,2}(\sqrt{2} + \sqrt{3}) = \sqrt{2} - \sqrt{3}.$$

On the other hand, $q^{\tau_2}(x) = x^2 + 2\sqrt{2}x - 1$, whose roots are $\gamma_{2,1} = -\sqrt{2} + \sqrt{3}$ and $\gamma_{2,2} = -\sqrt{2} - \sqrt{3}$. Thus τ_2 extends to two embeddings $\mathbb{Q}(\sqrt{2} + \sqrt{3}) \to \mathbb{C}$ defined by

$$\tau_{2,1}(\sqrt{2} + \sqrt{3}) = -\sqrt{2} + \sqrt{3},$$

$$\tau_{2,2}(\sqrt{2} + \sqrt{3}) = -\sqrt{2} - \sqrt{3}.$$

Proposition 4.12. *Let $L = K(\alpha)$, $n = [L : K]$, and let $g_1, g_2, \ldots, g_n$ denote the set of embeddings $g_i : L \to \mathbb{C}$. Let $\beta \in K(\alpha)$ and let $p(x)$ denote the irreducible polynomial of β over K. Then there exists an integer t for which*

$$p(x)^t = \prod_{i=1}^{n} (x - g_i(\beta)).$$

Proof. Let $s = [K(\beta) : K]$ and let $\beta = \beta_1, \beta_2, \ldots, \beta_s$ be the distinct roots of $p(x)$ giving rise to the set of embeddings $\tau_j : K(\beta) \to \mathbb{C}$ for $1 \le j \le s$. By Proposition 4.11, for each j, $1 \le j \le s$, there exists a set of $t = [K(\alpha) : K(\beta)]$ embeddings $\tau_{j,k}$, $1 \le k \le t$, that extend τ_j. For each i, $1 \le i \le n$, there exists a unique pair j, k with

$$g_i = \tau_{j,k}.$$

Now,

$$\begin{aligned}
\prod_{i=1}^{n} (x - g_i(\beta)) &= \prod_{j=1}^{s} \prod_{k=1}^{t} (x - \tau_{j,k}(\beta))) \\
&= \prod_{j=1}^{s} \prod_{k=1}^{t} (x - \tau_j(\beta)) \\
&= \prod_{k=1}^{t} \prod_{j=1}^{s} (x - \tau_j(\beta)) \\
&= \prod_{k=1}^{t} p(x) \\
&= p(x)^t.
\end{aligned}$$

□

As a consequence of Proposition 4.12 we have $\prod_{i=1}^{n} g_i(\beta) \in K$ and $\sum_{i=1}^{n} g_i(\beta) \in K$ for $\beta \in L$.

Definition 4.2. The map $\text{Norm}_{L/K} : L \to K$ defined as

$$\text{Norm}_{L/K}(\beta) = \prod_{i=1}^{n} g_i(\beta), \quad \beta \in L$$

is the **norm map of L over K**. The map $\text{tr}_{L/K} : L \to K$ defined as

$$\text{tr}_{L/K}(\beta) = \sum_{i=1}^{n} g_i(\beta), \quad \beta \in L$$

is the **trace map of L over K**.

4.2 Some Galois Theory

In this section we continue our assumption on K: $\mathbb{Q} \subseteq K \subseteq \mathbb{C}$, and we consider the splitting field L of an irreducible polynomial $p(x)$ over K. The splitting field L of $p(x)$ is the smallest field extension of K that contains all of the zeros of $p(x)$. Necessarily, L is a simple algebraic extension of K. The collection of all automorphisms of L that fix K is a group, called the Galois group of L/K, and to construct this group of automorphisms we will use the extension theorem of §4.1. We give some examples of Galois groups and state the Fundamental Theorem of Galois Theory.

* * *

Let K be a field $\mathbb{Q} \subseteq K \subseteq \mathbb{C}$ and let $p(x)$ be a monic irreducible polynomial of degree n over K. Then $p(x)$ has n distinct roots in $\mathbb{C}$ (Proposition 4.8) which we list as $\alpha_1, \alpha_2, \ldots, \alpha_n$. Now the finite extension

$$L = K(\alpha_1, \alpha_2, \ldots, \alpha_n),$$

is the smallest field extension of K that contains all of the zeros of $p(x)$. This is the **splitting field** L/K of $p(x)$.

A finite extension of fields L/K is a **Galois extension** if L is the splitting field of some monic irreducible polynomial $p(x)$ over K. If L/K is a Galois extension, then the set of all automorphisms of L that fix K is a group under function composition called the **Galois group** of L/K, denoted as $\text{Gal}(L/K)$. An **abelian extension** is a Galois extension with an abelian Galois group. Our definition of Galois extension is equivalent to that of other authors, who define a Galois extension L/K to be a normal, separable extension of K, that is, a Galois extension of K is the splitting field L/K of some family of polynomials over K which have distinct roots in some algebraic closure of K.

Proposition 4.13. *Let L/K be a Galois extension of fields with Galois group $G = Gal(L/K)$. Then $[L : K] = |G|$.*

Proof. By definition, L/K is the splitting field of some monic irreducible polynomial $p(x) \in K[x]$. Let $n_1 = \deg(p(x))$ and let β_1' be a root of $p_1(x) = p(x)$. Let $\tau_{i_1} : K(\beta_1') \to L$, $1 \leq i_1 \leq n_1$ be the set of n_1 embeddings of $K(\beta_1')$ into L that fix K. Let β_2 be some other root of $p(x)$ not in $K(\beta_1')$ (indeed, if such a root exists) and let β_2' be so that $K(\beta_2') = K(\beta_1')(\beta_2)$. Let $p_2(x)$ be the irreducible polynomial of β_2' over $K(\beta_1')$. Let $n_2 = \deg(p_2(x))$. By Proposition 4.11 each τ_{i_1} extends to n_2 embeddings $\tau_{i_1,i_2} : K(\beta_2') \to L$, $1 \leq i_2 \leq n_2$ that fix K.

Next, let β_3 be another root of $p(x)$ not in $K(\beta_2')$. Let β_3' be so that $K(\beta_3') = K(\beta_2')(\beta_3)$. Let $p_3(x)$ be the irreducible polynomial of β_3' over $K(\beta_2')$ with $n_3 = \deg(p_3(x))$. By Proposition 4.11 each τ_{i_1,i_2} extends to n_3 embeddings $\tau_{i_1,i_2,i_3} : K(\beta_3') \to L$, $1 \leq i_3 \leq n_3$ that fix K. Continuing in this manner we must have $K(\beta_q') = L$ for some $q \geq 1$. Thus there are $n_1 n_2 n_3 \cdots n_q = [L : K]$ automorphisms of L that fix K. □

Let L be the splitting field of the monic irreducible polynomial $p(x) \in K[x]$. By Proposition 4.9 there exists $\alpha \in L$ with $L = K(\alpha)$. Let $q(x) = \operatorname{irr}(\alpha, K)$ of degree $m = [L : K]$. Let $\alpha = \alpha_1, \alpha_2, \ldots, \alpha_m$ denote the roots of $q(x)$. As a consequence of Proposition 4.13 the m embeddings $g_i : L \to L$, $\alpha \mapsto \alpha_i$, $1 \leq i \leq m$, constitute the elements of $\operatorname{Gal}(L/K)$. These embeddings are permutations of the set of roots $\{\alpha_1, \alpha_2, \ldots, \alpha_m\}$, hence $\operatorname{Gal}(L/K)$ is a subgroup of S_m, the symmetric group on m letters. In what follows we compute some Galois groups.

Example 4.1. Let $p(x) = x^4 - 10x^2 + 1 \in \mathbb{Q}[x]$. We claim that $p(x)$ is an irreducible polynomial over $\mathbb{Q}$. To this end, suppose that $p(x)$ is reducible over $\mathbb{Q}$. By Gauss' Lemma this implies $p(x)$ is reducible over Z. If $p(x)$ has a linear factor over Z, then necessarily, ± 1 is a root of $p(x)$, which is not true. Thus $p(x)$ factors into two polynomials each of degree 2:

$$p(x) = x^4 - 10x^2 + 1 = (ax^2 + bx + c)(dx^2 + ex + f),$$

for some $a, b, c, d, e, f \in Z$. But then

$$\begin{cases} ad = 1 \\ ae + bd = 0 \\ af + be + cd = -10 \\ bf + ce = 0 \\ cf = 1, \end{cases}$$

whence either $e^2 = 8$, or $e^2 = 12$, both impossibilities.

Now, the zeros of $p(x)$ are:

$$\alpha_1 = \sqrt{2} + \sqrt{3},\ \alpha_2 = -\sqrt{2} - \sqrt{3},\ \alpha_3 = -\sqrt{2} + \sqrt{3},\ \alpha_4 = \sqrt{2} - \sqrt{3},$$

and the splitting field K of $p(x)$ over $\mathbb{Q}$ is $\mathbb{Q}(\sqrt{2} + \sqrt{3})$. Thus $[K : \mathbb{Q}] = 4 = |\mathrm{Gal}(K/\mathbb{Q})|$. And so, $\mathrm{Gal}(K/\mathbb{Q})$ is either isomorphic to Z_4 or $Z_2 \times Z_2$, cf. §1.4.

We can be more precise: The embeddings $\tau_{1,1}, \tau_{1,2}, \tau_{2,1}, \tau_{2,2}$ (§4.1) constitute the elements of $\mathrm{Gal}(K/\mathbb{Q})$ and so, $\mathrm{Gal}(K/Q) \cong Z_2 \times Z_2$. As a subgroup of S_4 (with the roots given by their subscripts) $\mathrm{Gal}(K/\mathbb{Q})$ consists of

$$\begin{pmatrix} 1\,2\,3\,4 \\ 1\,2\,3\,4 \end{pmatrix}, \begin{pmatrix} 1\,2\,3\,4 \\ 4\,2\,3\,1 \end{pmatrix}, \begin{pmatrix} 1\,2\,3\,4 \\ 3\,2\,1\,4 \end{pmatrix}, \begin{pmatrix} 1\,2\,3\,4 \\ 2\,1\,3\,4 \end{pmatrix}.$$

Example 4.2. Let $K = \mathbb{Q}(\sqrt{2})$. The polynomial $p(x) = x^2 - 2\sqrt{2}x - 1$ is irreducible over K. (Why?) The roots of $p(x)$ are $a_1 = \sqrt{2} + \sqrt{3}$, $a_2 = \sqrt{2} - \sqrt{3}$, and so the splitting field of $p(x)$ is $L = K(\sqrt{2} + \sqrt{3})$. One has $\mathrm{Gal}(L/K) = \{\tau_{1,1}, \tau_{1,2}\} \cong Z_2$.

Example 4.3. Let p be a prime number. The polynomial

$$p(x) = x^{p-1} + x^{p-2} + \cdots + x^3 + x^2 + x + 1$$

is irreducible over $\mathbb{Q}$ (Proposition 2.8). The roots of $p(x)$ are

$$\alpha_1 = \zeta_p,\ \alpha_2 = \zeta_p^2,\ \alpha_3 = \zeta_p^3, \ldots, \alpha_{p-1} = \zeta_p^{p-1}$$

(§2.8, Exercise 31). Thus the splitting field is $K = \mathbb{Q}(\zeta_p)$ and so $|\mathrm{Gal}(K/\mathbb{Q})| = p - 1$. In fact, the embeddings are given as $\tau_i : K \to K$, $\zeta_p \mapsto \zeta_p^i$, $1 \leq i \leq p - 1$. If r is a primitive root modulo p, then $\mathrm{Gal}(K/L) = \langle \tau_r \rangle \cong U(Z_p) \cong Z_{p-1}$.

Example 4.4. Let $p(x) = x^3 - 2$. By the Eisenstein Criterion, $p(x) = x^3 - 2$ is irreducible over $\mathbb{Q}$. The roots of $p(x)$ are $\sqrt[3]{2}$, $\zeta_3\sqrt[3]{2}$, $\zeta_3^2\sqrt[3]{2}$, and so, the splitting field of $p(x)$ over $\mathbb{Q}$ is $K = \mathbb{Q}(\zeta_3, \sqrt[3]{2})$. By Proposition 4.9, $K = \mathbb{Q}(\zeta_3 + \sqrt[3]{2})$. Observe that $\mathbb{Q} \subseteq \mathbb{Q}(\zeta_3) \subseteq K$. We shall compute

$\text{Gal}(K/\mathbb{Q})$ using the Extension Theorem (Proposition 4.11). There are two automorphisms of $\mathbb{Q}(\zeta_3)$ that fix $\mathbb{Q}$ defined by

$$\tau_1(\zeta_3) = \zeta_3,$$

$$\tau_2(\zeta_3) = \zeta_3^2.$$

Now the irreducible polynomial of $\zeta_3 + \sqrt[3]{2}$ over $\mathbb{Q}(\zeta_3)$ is

$$q(x) = x^3 - 3\zeta_3 x^2 + 3\zeta_3^2 x - 3.$$

We have

$$q^{\tau_1}(x) = x^3 - 3\zeta_3 x^2 + 3\zeta_3^2 x - 3.$$

The roots of $q^{\tau_1}(x)$ are

$$\zeta_3 + \sqrt[3]{2}, \quad \zeta_3 + \zeta_3\sqrt[3]{2}, \quad \zeta_3 + \zeta_3^2\sqrt[3]{2}.$$

Thus, τ_1 extends to three automorphisms of K that fix $\mathbb{Q}$ defined by

$$\tau_{1,1}(\zeta_3 + \sqrt[3]{2}) = \zeta_3 + \sqrt[3]{2},$$

$$\tau_{1,2}(\zeta_3 + \sqrt[3]{2}) = \zeta_3 + \zeta_3\sqrt[3]{2},$$

$$\tau_{1,3}(\zeta_3 + \sqrt[3]{2}) = \zeta_3 + \zeta_3^2\sqrt[3]{2}.$$

Moreover,

$$q^{\tau_2}(x) = x^3 - 3\zeta_3^2 x^2 + 3\zeta_3 x - 3,$$

with roots

$$\zeta_3^2 + \sqrt[3]{2}, \quad \zeta_3^2 + \zeta_3\sqrt[3]{2}, \quad \zeta_3^2 + \zeta_3^2\sqrt[3]{2}.$$

Thus, τ_2 extends to three automorphisms of K that fix $\mathbb{Q}$ defined by

$$\tau_{2,1}(\zeta_3 + \sqrt[3]{2}) = \zeta_3^2 + \sqrt[3]{2},$$

$$\tau_{2,2}(\zeta_3 + \sqrt[3]{2}) = \zeta_3^2 + \zeta_3\sqrt[3]{2},$$

$$\tau_{2,3}(\zeta_3 + \sqrt[3]{2}) = \zeta_3^2 + \zeta_3^2\sqrt[3]{2}.$$

The six elements $\{\tau_{1,1}, \tau_{1,2}, \tau_{1,3}, \tau_{2,1}, \tau_{2,2}, \tau_{2,3}\}$ constitute $G = \text{Gal}(K/\mathbb{Q})$. Since $\text{Gal}(K/\mathbb{Q}) \leq S_3$, we have $G \cong S_3$.

Example 4.5. Let p be a prime number and let a be a positive integer that is not a pth power in Z. Then $p(x) = x^p - a$ is irreducible over $\mathbb{Q}$. The roots of $p(x)$ are

$$\sqrt[p]{a},\ \zeta_p \sqrt[p]{a},\ \zeta_p^2 \sqrt[p]{a},\ \dots\ \zeta_p^{p-1} \sqrt[p]{a}.$$

Put $K = \mathbb{Q}(\zeta_p)$. The splitting field of $p(x)$ over K is $L = K(\sqrt[p]{a})$. One has

$$G = \text{Gal}(L/K) = \{\tau_i : L \to L,\ \tau_i(\sqrt[p]{a}) = \zeta_p^i \sqrt[p]{a}\},$$

$0 \le i \le p-1$. Thus $G \cong Z_p$.

We state the fundamental theorem of Galois theory.

Proposition 4.14 (Fundamental Theorem of Galois Theory). *Suppose L/K is a Galois extension of fields with Galois group $G =$ Gal(L/K).*

(i) If H is a subgroup of G, then

$$L^H = \{x \in L : \ h(x) = x, \forall h \in H\}$$

is a subfield of L. Moreover, L is a Galois extension of L^H with Gal$(L/L^H) \cong H$. If H is a normal subgroup of G, then L^H is a Galois extension of K with Gal$(L^H/K) \cong G/H$.

(ii) Let E, $K \subseteq E \subseteq L$ be an intermediate field. Then

$$H_E = \{g \in G : \ g(x) = x, \forall x \in E\}$$

is a subgroup of G, and L is a Galois extension of E with group H_E. If E is a Galois extension of K then H_E is a normal subgroup of G.

(iii) Let $\mathcal{S}(G)$ denote the collection of subgroups of G, and let $\mathcal{F}(L/K)$ denote the collection of intermediate fields between K and L. Then there is an order-reversing bijective map

$$\Psi : \mathcal{S}(G) \to \mathcal{F}(L/K),$$

defined as $H \mapsto L^H$.

Proof. For (i): Suppose $H \le G$. Clearly, $K \subseteq L^H$. Let $x, y \in L^H$, and let $h \in H$. Then $h(x+y) = h(x) + h(y) = x + y$, and $h(xy) = h(x)h(y) = xy$, and so, L^H is a subring of L. Consequently, L^H is a subfield of L. We claim that L/L^H is a Galois extension. To this end, write $L = K(\alpha)$ where α is a root of an irreducible monic polynomial $p(x)$ over K. By Proposition 4.12,

$$p(x) = \prod_{g \in G} (x - g(\alpha)).$$

Put $q(x) = \prod_{h \in H}(x - h(\alpha))$. Since each $h \in H$ permutes the set $\{h(\alpha)\}_{h \in H}$, the coefficients of $q(x)$ are fixed by H. Thus $q(x) \in L^H[x]$. Since one of the factors of $q(x)$ is $x - e(\alpha) = x - \alpha$, $q(\alpha) = 0$. Suppose there is a polynomial $r(x)$ over L^H of degree smaller than $|H| = \deg(q(x))$ with $r(\alpha) = 0$. Then the set $\{h(\alpha)\}_{h \in H}$ constitutes distinct zeros of $r(x)$, contradicting Proposition 2.6. Thus $q(x)$ is irreducible over L^H. Since L is the splitting field of $q(x)$, L/L^H is Galois. Moreover, $\mathrm{Gal}(L/L^H) = H$.

Now assume that $H \triangleleft G$. By Proposition 4.9, $L^H = K(\eta)$ for some $\eta \in L^H$. Observe that $g^{-1}hg \in H$ for all $h \in H$, $g \in G$. Thus for $a \in L^H$, $(g^{-1}hg)(a) = a$ and so, $h(g(a)) = g(a)$ for all $h \in H$. Thus $g(a) \in L^H$ for all $g \in G$, and so, $g(\eta) \in L^H$ for all $g \in G$. It follows that L^H is the splitting field of the irreducible polynomial of η. Thus L^H/K is a Galois extension of fields. Now, let gH be an element in the factor group G/H, and let $gh \in gH$. Now for $a \in L^H$, $(gh)(a) = g(h(a)) = g(a)$, and so, the Galois action of elements in the coset gH depends only on the action of the representative g. It follows that $\mathrm{Gal}(L^H/K) = G/H$.

For (ii): Let $h, h' \in H_E$, $x \in E$. Then $(hh')(x) = h(h'(x)) = h(x) = x$, hence H_E is closed under the binary operation of G. Moreover, $e(x) = x$, so that $e \in H_E$. Lastly, $x = e(x) = (h^{-1}h)(x) = h^{-1}(h(x)) = h^{-1}(x)$, so that $h^{-1} \in H_E$. Thus $H_E \leq G$.

Clearly, $E \subseteq L^{H_E}$. Let $\alpha \in L \backslash E$. By Proposition 4.11, there is an automorphism $L \to L$ fixing E which does not fix α. So $\alpha \notin L^{H_E}$. Thus $L^{H_E} \subseteq E$, and so $E = L^{H_E}$. Hence, by (i) L/E is Galois with group H_E.

Now suppose that E/K is a Galois extension with group G'. Define a map $\phi : G \to G'$ by the rule $\phi(g)(x) = g(x)$ for $g \in G$, $x \in E$. Every element $g \in G$ is the extension to L of some element $g' \in G'$. Thus for each $g \in G$, there exists an element $g' \in G'$ so that $g(x) = g'(x)$ for $x \in E$. Thus ϕ is surjective. Now, for $x \in E$,

$$\begin{aligned}
\phi(g_1 g_2)(x) &= (g_1 g_2)(x) \\
&= g_1(g_2(x)) \\
&= g_1(g_2'(x)) \\
&= g_1'(g_2'(x)) \\
&= \phi(g_1)(\phi(g_2)(x)) \\
&= (\phi(g_1)\phi(g_2))(x)
\end{aligned}$$

and so, ϕ is a group homomorphism. If follows that $\ker(\phi) \triangleleft G$. But $\ker(\phi) = H_E$ and so the proof is complete.

For (iii): Suppose $H' \leq H \leq G$. Then $L^H \subseteq L^{H'}$, hence Ψ is order-reversing. Define a map $\Theta : \mathcal{F}(L/K) \to \mathcal{S}(G)$ by the rule $E \mapsto H_E$. We show that $L^{H_E} = E$ and $H_{L^H} = H$, thus $\Psi\Theta$ is the identity map on $\mathcal{F}(L/K)$ and $\Theta\Psi$ is the identity on $\mathcal{S}(G)$, thus Ψ is a bijection. From (ii) we already know that $L^{H_E} = E$. We show that $H = H_{L^H}$. Since H_{L^H} is the subgroup of G leaving L^H fixed, (ii) implies that L/L^H is Galois with group H_{L^H}. By (i), L/L^H is Galois with group H, hence $H = H_{L^H}$. □

In the following example we illustrate the fundamental theorem.

Example 4.6. Let ζ_7 denote a primitive 7th root of unity and consider the Galois extension $K = \mathbb{Q}(\zeta_7)$ over $\mathbb{Q}$ with Galois group $G = Z_7^*$. Since 3 is a primitive root modulo 7, G is generated by $\tau_3 : K \to K$, $\zeta_7 \mapsto \zeta_7^3$. Now, $\tau_3^3 : K \to K$, $\zeta_7 \mapsto \zeta_7^6$ generates a normal subgroup $H = \langle \tau_3^3 \rangle$ of order 2. The fixed field of H is

$$K^H = \{x \in K : \ \tau_3^3(x) = x\} = \mathbb{Q}(\zeta_7 + \zeta_7^6).$$

Now, K Galois over K^H; K is the splitting field of the monic irreducible polynomial

$$p(x) = x^2 - x(\zeta_7 + \zeta_7^6) + 1$$

over K^H and $\mathrm{Gal}(K/K^H) = H$. Moreover, K^H is Galois over $\mathbb{Q}$; K^H is the splitting field of the monic irreducible polynomial

$$q(X) = x^3 + x^2 - 2x - 1$$

over $\mathbb{Q}$ and $\mathrm{Gal}(K^H/\mathbb{Q}) = \langle \tau_3 \rangle / H \cong Z_3$.

4.3 The Ring of Integers

In this section we define the ring of integers R of a simple algebraic extension $K = \mathbb{Q}(\alpha)$. We show that every non-zero ideal of R contains a $\mathbb{Q}$-basis for K, specifically, R is a free Z-module of rank $n = [K : \mathbb{Q}]$. We show how to construct the ring of integers of $\mathbb{Q}(\sqrt{m})$ and compute the ring of integers of $\mathbb{Q}(\zeta_p)$, where ζ_p is a primitive pth root of unity. We employ Dirichlet's Unit Theorem to construct the group of units of the ring of integers R. Specifically, we compute $U(R)$ where R is the ring of integers of a quadratic field extension and where R is the ring of integers of $\mathbb{Q}(\zeta_p)$.

* * *

Let $K = \mathbb{Q}(\alpha)$ be a simple algebraic extension of $\mathbb{Q}$ of degree n. By Proposition 4.3 every element in K is the zero of a monic polynomial of degree $\leq n$ with coefficients in $\mathbb{Q}$. If we restrict coefficients to Z, we determine a special subset of K: An element $\beta \in K$ which is a zero of a monic polynomial with coefficients in Z is **integral** over Z. The collection of all elements of K which are integral over Z is the **integral closure of Z in K**.

We want to prove that the integral closure of Z in K is a ring and to do this we need a general fact.

Proposition 4.15. *Let S be a commutative ring that is also a finitely generated module over a commutative ring with unity R. Let $S \subseteq T$ be an inclusion of rings with T a commutative ring with unity. Assume that $R \subseteq T$ and that the multiplication RS is the scalar multiplication on S. Assume that for $x \in T$, $xS = 0$ implies that $x = 0$. Let α be an element of T for which $\alpha S \subseteq S$. Then α is a root of a monic polynomial over R.*

Proof. Let $\{b_1, b_2, \ldots, b_m\}$ be a generating set for S over R. There exist elements $r_{i,j} \in R$ for which

$$\begin{aligned} \alpha b_1 &= r_{1,1}b_1 + r_{1,2}b_2 + \cdots + r_{1,m}b_m, \\ \alpha b_2 &= r_{2,1}b_1 + r_{2,2}b_2 + \cdots + r_{2,m}b_m, \\ &\vdots \\ \alpha b_m &= r_{m,1}b_1 + r_{m,2}b_2 + \cdots + r_{m,m}b_m. \end{aligned}$$

Thus,

$$\begin{aligned} (\alpha - r_{1,1})b_1 - r_{1,2}b_2 - \cdots - r_{1,m}b_m &= 0, \\ -r_{2,1}b_1 + (\alpha - r_{2,2})b_2 - \cdots - r_{2,m}b_m &= 0, \\ &\vdots \\ -r_{m,1}b_1 - r_{m,2}b_2 - \cdots + (\alpha - r_{m,m})b_m &= 0. \end{aligned}$$

Put $C = (r_{i,j}) \in \mathrm{Mat}_m(R)$. Then $(\alpha I_m - C)B = 0$, where $B = (b_1, b_2, \ldots, b_m)^T$. Now,

$$0 = (\mathrm{adj}(\alpha I_m - C))(\alpha I_m - C)B = \det(\alpha I_m - C)I_m B.$$

It follows that $\det(\alpha I_m - C)S = 0$, and so, $\det(\alpha I_m - C) = 0$. Hence α is a zero of the monic polynomial $f(x) = \det(xI_m - C) \in R[x]$ of degree m. □

Proposition 4.16. *Let K be a simple algebraic extension of $\mathbb{Q}$. Then the integral closure of Z in K is closed under the addition and multiplication of K.*

Proof. Let γ, β be integral over Z. We show that $\gamma\beta$ is integral over Z. Since γ is integral over Z, there exists a monic polynomial $f(x) \in Z[x]$ of degree k for which $f(\gamma) = 0$. Likewise, there exists a monic polynomial $g(x) \in Z[x]$ of degree n for which $g(\beta) = 0$.

Now let A be the Z-module generated by $\{1, \gamma, \gamma^2, \dots, \gamma^{k-1}\}$ and let B be the Z-module generated by $\{1, \beta, \beta^2, \dots, \beta^{n-1}\}$. Let AB be the Z-module generated by $\{\gamma^i\beta^j\}$, $0 \le i \le k-1$, $0 \le j \le n-1$. Then $\gamma\beta AB \subseteq AB$. Now, $AB \subseteq K$ is an inclusion of rings with the following properties: for $x \in K$, $xAB = 0$ implies $x = 0$, and AB is a finitely generated Z-module. Thus Proposition 4.15 applies to show that $\gamma\beta$ is integral over Z.

A similar argument shows that $\gamma+\beta$ is integral over Z (see §4.8, Exercise 15). □

It follows from Proposition 4.16 that the set of elements of K that are integral over Z is a ring. We define this ring to be **the ring of integers of K**. We shall usually denote the ring of integers of K as R. The simplest example of a ring of integers is Z: the integral closure of Z in $\mathbb{Q}$ is Z (§4.8, Exercise 16).

Let $\beta \in K$ be integral over Z, that is, suppose

$$a_0 + a_1\beta + a_2\beta^2 + \cdots + a_{m-1}\beta^{m-1} + \beta^m = 0,$$

for some monic polynomial $f(x) = a_0 + a_1x + \cdots a_{m-1}x^{m-1} + x^m$ over Z. Now for each embedding $g_i : K \to \mathbb{C}$,

$$\begin{aligned} 0 &= g_i(a_0 + a_1\beta + a_2\beta^2 + \cdots + a_{m-1}\beta^{m-1} + \beta^m) \\ &= a_0 + a_1g_i(\beta) + a_2g_i(\beta)^2 + \cdots + a_{m-1}g_i(\beta)^{m-1} + g_i(\beta)^m, \end{aligned}$$

so that $g_i(\beta)$ is integral over Z for all i.

Proposition 4.17. *The norm and trace maps $Norm_{K/\mathbb{Q}}$ and $Tr_{K/\mathbb{Q}}$ restrict to the maps $Norm_{K/\mathbb{Q}} : R \to Z$ and $Tr_{K/\mathbb{Q}} : R \to Z$.*

Proof. Let $\beta \in R$. Since $g_i(\beta)$ is integral over Z for all i, $\mathrm{Norm}_{L/K}(\beta) = \prod_{i=1}^n g_i(\beta)$ and $\mathrm{Tr}_{L/K}(\beta) = \sum_{i=1}^n g_i(\beta)$ are integral over Z by Proposition 4.16. Since these quantities are also in $\mathbb{Q}$, we conclude that $\mathrm{Norm}_{K/\mathbb{Q}}(\beta)$ and $\mathrm{Tr}_{K/\mathbb{Q}}(\beta)$ are integers. □

If $\beta \in K$ is integral over Z, then irreducible polynomial of β has coefficients in Z.

Proposition 4.18. *Let $K = \mathbb{Q}(\alpha)$, $n = [K : \mathbb{Q}]$ and let R denote the ring of integers of K. Let $\beta \in R$ and let $p(x)$ be the irreducible polynomial of β over $\mathbb{Q}$. Then $p(x) \in Z[x]$.*

Proof. Since β is integral over Z, there exists a monic polynomial $q(x) \in Z[x]$ with β as a root. Now, $p(x)$ divides $q(x)$ in $\mathbb{Q}[x]$, and so, $q(x) = p(x)s(x)$ for some $s(x)$ in $\mathbb{Q}[x]$. By Gauss' Lemma, there exist polynomials $\hat{p}(x)$, $\hat{s}(x)$ over Z, associates of $p(x)$ and $s(x)$ respectively, with $q(x) = \hat{p}(x)\hat{s}(x)$. Comparing leading coefficients, $\hat{p}(x)$ and $\hat{s}(x)$ must both be monic or the negatives of monic polynomials. Thus $p(x) = \pm\hat{p}(x)$, and hence is in $Z[x]$. □

Lemma 4.1. *Let $K = \mathbb{Q}(\alpha)$, $n = [K : \mathbb{Q}]$ and let R denote the ring of integers of K. Every non-zero ideal J of R contains a basis for K over $\mathbb{Q}$.*

Proof. Let $\beta \in K = \mathbb{Q}(\alpha)$. Then the set $\{1, \beta, \beta^2, \ldots, \beta^n\}$ is linearly dependent over $\mathbb{Q}$, and so,

$$a_0 + a_1\beta + a_2\beta^2 + \cdots + a_n\beta^n = 0,$$

for integers $a_i \in Z$, not all 0. Let j be the largest index for which $a_j \neq 0$. Then

$$a_j^{j-1}(a_0 + a_1\beta + a_2\beta^2 + \cdots + a_j\beta^j) = 0,$$

and so,

$$a_j^{j-1}a_0 + a_j^{j-2}a_1(a_j\beta) + a_j^{j-3}a_2(a_j\beta)^2 + \cdots + (a_j\beta)^j = 0,$$

thus, $a_j\beta$ is integral over Z. Hence $a_j\beta \in R$.

Next, let $\{b_1, b_2, \ldots, b_n\}$ be a basis for K over $\mathbb{Q}$. Now by the preceding paragraph there exist integers c_i for which $\{c_ib_i\}_{i=1}^n \subseteq R$. Let a be any non-zero element of J. Then $\{c_ib_ia\} \subseteq J$ is a basis for $K/\mathbb{Q}$. □

Proposition 4.19. *Let $K = \mathbb{Q}(\alpha)$, $n = [K : \mathbb{Q}]$ and let R denote the ring of integers of K. Then R is an integral domain with Frac(R) $= K$.*

Proof. Clearly, R is an integral domain. Let $\phi : R \to K$ be the inclusion map and let $\lambda : R \to \text{Frac}(R)$ be the localization map. By the UMPL

(Propositon 2.45) there exists a unique inclusion ψ : Frac$(R) \to K$. Thus Frac$(R) \subseteq K$.

Since $Z \subseteq R$, there exists an inclusion $\phi' : Z \to$ Frac(R). Let $\lambda' : Z \to \mathbb{Q}$ denote the localization map. By the UMPL, there is a unique inclusion $\psi' : \mathbb{Q} \to$ Frac(R). Since R contains a $\mathbb{Q}$-basis for K (Lemma 4.1), $K \subseteq$ Frac(R). □

Proposition 4.20. *Let K be a simple algebraic extension of $\mathbb{Q}$ with ring of integers R. Then $K = \mathbb{Q}(\alpha)$ for some $\alpha \in R$.*

Proof. Write $K = \mathbb{Q}(\beta)$ for some $\beta \in K$. Now by Proposition 4.19, there exists a non-zero element $r \in R$ with $r\beta \in R$. Now $K = \mathbb{Q}(\alpha)$ with $\alpha = r\beta \in R$. □

Proposition 4.20 says that a simple algebraic extension of $\mathbb{Q}$ can be written in the form $K = \mathbb{Q}(\alpha)$ where $\alpha \in R$, the ring of integers of K. Suppose that $[K : \mathbb{Q}] = n$. By Proposition 4.18, α is a zero of a monic polynomial of degree n with coefficients in Z. It follows that

$$Z[\alpha] = Z \oplus Z\alpha \oplus Z\alpha^2 \oplus \cdots \oplus Z\alpha^{n-1}$$

is contained in R. It may be, however, that R is larger than $Z[\alpha]$, and in certain cases, we can compute the ring of integers precisely. Let m be a square-free integer and let $K = \mathbb{Q}(\sqrt{m})$. We want to determine which elements of $\mathbb{Q}(\sqrt{m})$ are integral over Z. Note that any $\alpha \in \mathbb{Q}(\sqrt{m})$ can be written in the form

$$\alpha = \frac{a + b\sqrt{m}}{c}$$

where a, b, c are integers with $\gcd(a, b, c) = 1$.

Lemma 4.2. *If α, as above, is integral over Z and $\alpha \neq 0$, then either $c = 1$ or $c = 2$.*

Proof. Note that the irreducible polynomial of α is

$$p(x) = x^2 - \frac{2a}{c}x + \frac{a^2 - b^2m}{c^2}.$$

Thus by Proposition 4.18, $p(x) \in Z[x]$, that is, $\frac{2a}{c}$ and $\frac{a^2-b^2m}{c^2}$ are in Z.

Let $p > 2$ be a prime number. Then $p \nmid c$. To see this suppose that $p \mid c$. Then since $c \mid 2a$, $p \mid a$. Now since $c^2 \mid (a^2 - b^2m)$ and m is square-free, $p \mid b$. But this says that $\gcd(a, b, c) \neq 1$, a contradiction. So, $p \nmid c$, and hence $c = 2^j$ for some integer $j \geq 0$. Suppose that $j \geq 2$. Then $2^{j-1} \mid a$, $2^{j-1} \mid b$ (again because m is square-free), and thus $\gcd(a, b, c) \neq 1$. Conclusion: either $c = 1$ or $c = 2$. □

Proposition 4.21. *Let $K = \mathbb{Q}(\sqrt{m})$, where m is a square-free integer. Then:*

(i) $R = \mathbb{Z}[\sqrt{m}]$ if $m \equiv 2, 3 \bmod 4$;

(ii) $R = Z[\frac{1+\sqrt{m}}{2}]$ if $m \equiv 1 \bmod 4$.

Proof. For (i), we already know that $Z[\sqrt{m}] \subseteq R$, so the problem is to prove the reverse containment. Let $\alpha \in R$. Then by Lemma 4.2, $\alpha = \frac{a+b\sqrt{m}}{c}$ with $\gcd(a,b,c) = 1$ and either $c = 1$ or $c = 2$. If $c = 1$, then certainly, $\alpha \in Z[\sqrt{m}]$, and thus $R = Z[\sqrt{m}]$. So we assume that $c = 2$. From the proof of Lemma 4.2, we see that $4 \mid (a^2 - b^2m)$, hence $a^2 \equiv b^2m \bmod 4$. Consequently, either both a^2 and b^2 are congruent to 0 modulo 4, or both a^2 and b^2 are congruent to 1 modulo 4. In either case, $a \equiv b \bmod 2$.

If $a^2, b^2 \equiv 0 \bmod 4$, then $\gcd(a,b,c) \neq 1$, so we conclude that $a^2, b^2 \equiv 1 \bmod 4$, whence $1 \equiv m \bmod 4$. Since we've assumed that $m \equiv 2, 3 \bmod 4$, we cannot have $c = 2$.

For (ii), the condition $m \equiv 1 \bmod 4$ implies that $\frac{1+\sqrt{m}}{2} \in R$, and hence $Z[\frac{1+\sqrt{m}}{2}] \subseteq R$. Now suppose that $\alpha \in R$. By Lemma 4.2, $\alpha = \frac{a+b\sqrt{m}}{c}$ with $\gcd(a,b,c) = 1$ and either $c = 1$ or $c = 2$. Clearly $\alpha \in Z[\frac{1+\sqrt{m}}{2}]$ if $c = 1$.

If $c = 2$ then from (i) we see that $\alpha = \frac{a+b\sqrt{2}}{2}$ with $a \equiv b \bmod 2$. Now,

$$\frac{a + b\sqrt{m}}{2} = a\left(\frac{1+\sqrt{m}}{2}\right) + \frac{b-a}{2}\sqrt{m},$$

which is in $Z[\frac{1+\sqrt{m}}{2}]$ since $a \equiv b \bmod 2$. Thus $R \subseteq Z[\frac{1+\sqrt{m}}{2}]$. □

For example, if $K = \mathbb{Q}(\mathrm{i})$, then $R = \mathbb{Z}[\mathrm{i}]$.

In the case that $m \equiv 1 \bmod 4$, note that $\sqrt{m} \in R$, yet $Z[\sqrt{m}]$ is not all of R. For instance, in the case $m = -23$, the ring of integers of the imaginary quadratic field extension $K = \mathbb{Q}(\sqrt{-23})$ is $R = Z[(1+\sqrt{-23})/2]$. (Recall that R is the ring given in §3.2 in the illustration of Proposition 3.12.) By Proposition 4.19, R is an integral domain with $\mathrm{Frac}(R) = \mathbb{Q}(\sqrt{-23})$. As noted in §3.2, R is not a PID, indeed the ideal $T = (3, 2+\sqrt{-23})$ is not principal. To prove this, we employ the norm map $\mathrm{Norm}_{K/\mathbb{Q}} : R \to Z$, which as one can check is given as

$$\begin{aligned}\mathrm{Norm}_{K/\mathbb{Q}}(a + b(1+\sqrt{-23})/2) &= (a + b(1+\sqrt{-23})/2)(a + b(1-\sqrt{-23})/2)\\ &= a^2 + ab + 6b^2,\end{aligned}$$

for $a, b \in Z$. By way of contradiction, we assume that $T = (r)$ for some $r \in R$. Since $3 \in T$, $3 = rs$ for some $s \in R$. Applying the norm map yields

$$\mathrm{Norm}_{K/\mathbb{Q}}(rs) = \mathrm{Norm}_{K/\mathbb{Q}}(r)\mathrm{Norm}_{K/\mathbb{Q}}(s) = 9.$$

Thus $\mathrm{Norm}_{K/\mathbb{Q}}(r) \mid 9$. Moreover, $1 - \sqrt{-23} \in T$, and so there exists $t \in R$ with $1 - \sqrt{-23} = rt$. An application of the norm map yields

$$\mathrm{Norm}_{K/\mathbb{Q}}(r)\mathrm{Norm}_{K/\mathbb{Q}}(t) = (1 - \sqrt{-23})(1 + \sqrt{-23}) = 24,$$

and so, $\mathrm{Norm}_{K/\mathbb{Q}}(r) = 3$. This is impossible since the norm of any element in R other than ± 1 is at least 6. Thus T is not principal and R is not a PID.

Let p be a prime and let $K = \mathbb{Q}(\zeta_p)$. We want to compute the ring of integers R of K. We first prove two lemmas.

Lemma 4.3. *$R(1 - \zeta_p) \cap Z = pZ$.*

Proof. From the formula $p = \prod_{i=1}^{p-1}(1 - \zeta_p^i)$ one obtains $p = r(1 - \zeta_p)$ with $r = \prod_{i=2}^{p-1}(1 - \zeta_p^i) \in R$. Thus $pZ \subseteq R(1 - \zeta_p) \cap Z$. Note that $R(1 - \zeta_p) \cap Z$ is an ideal of Z containing the maximal ideal pZ. Thus either $pZ = R(1 - \zeta_p) \cap Z$ or $R(1 - \zeta_p) \cap Z = Z$. In the latter case, there exists an element $\alpha \in R$ with $\alpha(1 - \zeta_p) = 1$. Applying the norm map yields

$$\begin{aligned} 1 &= \mathrm{Norm}_{K/\mathbb{Q}}(\alpha(1 - \zeta_p)) \\ &= \mathrm{Norm}_{K/\mathbb{Q}}(\alpha)\mathrm{Norm}_{K/\mathbb{Q}}(1 - \zeta_p) \\ &= \mathrm{Norm}_{K/\mathbb{Q}}(\alpha) \prod_{i=1}^{p-1}(1 - \zeta_p^i) \\ &= p \cdot \mathrm{Norm}_{K/\mathbb{Q}}(\alpha), \end{aligned}$$

where $\mathrm{Norm}_{K/\mathbb{Q}}(\alpha)$ is an integer. This is impossible. Hence $pZ = R(1 - \zeta_p) \cap Z$. $\square$

Lemma 4.4. *$\mathrm{Tr}_{K/\mathbb{Q}}(R(1 - \zeta_p)) \subseteq pZ$.*

Proof. Let $\alpha \in R$. Then

$$\begin{aligned} \mathrm{Tr}_{K/\mathbb{Q}}(\alpha(1-\zeta_p)) &= g_1(\alpha(1-\zeta_p)) + g_2(\alpha(1-\zeta_p)) + \cdots + g_{p-1}(\alpha(1-\zeta_p)) \\ &= g_1(\alpha)(1-\zeta_p) + g_2(\alpha)(1-\zeta_p^2) + \cdots + g_{p-1}(\alpha)(1-\zeta_p^{p-1}) \\ &\in R(1-\zeta_p), \end{aligned}$$

since $g_i(\alpha)$ are integral elements. Since $\mathrm{Tr}_{K/\mathbb{Q}}(\alpha(1-\zeta_p))$ is an integer, one has $\mathrm{Tr}_{K/\mathbb{Q}}(\alpha(1-\zeta_p)) \in R(1-\zeta_p) \cap Z$. Hence $\mathrm{Tr}_{K/\mathbb{Q}}(\alpha(1-\zeta_p)) \in pZ$ by Lemma 4.3. □

Now we can compute the ring of integers of $K = \mathbb{Q}(\zeta_p)$.

Proposition 4.22. *Let p be a prime number and let $K = \mathbb{Q}(\zeta_p)$. Then $R = Z[\zeta_p]$.*

Proof. Since $\zeta_p \in R$, we easily have $Z[\zeta_p] \subseteq R$. The problem is to show the reverse containment. Let $\alpha \in R$ and write

$$\alpha = a_0 + a_1\zeta_p + a_2\zeta_p^2 + \cdots + a_{p-2}\zeta_p^{p-2},$$

with $a_0, a_1, \ldots, a_{p-2} \in \mathbb{Q}$. For each i, $0 \le i \le p-2$,

$$\alpha\zeta_p^{-i}(1-\zeta_p) = a_0\zeta^{-i}(1-\zeta_p) + a_1\zeta^{1-i}(1-\zeta_p) + \cdots + a_i(1-\zeta_p) + \cdots + a_{p-2}\zeta_p^{p-2-i}(1-\zeta_p)$$

is an element of $R(1-\zeta_p)$. Applying the trace map yields

$$\mathrm{Tr}_{K/\mathbb{Q}}(\alpha\zeta_p^{-i}(1-\zeta_p)) = pa_i.$$

By Lemma 4.4, $pa_i \in pZ$, and so $a_i \in Z$.

For a different proof, see [Underwood (2011), Proposition 10.2.15]. □

In fact, we can use the methods of Proposition 4.22 to extend the result.

Proposition 4.23. *Let p be a prime number, let $n \ge 1$ be an integer and let $K = \mathbb{Q}(\zeta_{p^n})$. Then $R = Z[\zeta_{p^n}]$.*

Proof. See §4.8, Exercise 18. □

Let $K = \mathbb{Q}(\alpha)$ be a simple algebraic extension of $\mathbb{Q}$ of degree n. Let R denote the ring of integers in K. We are interested in computing units of R. In the case $K = \mathbb{Q}(\zeta_p)$, $R = Z[\zeta_p]$, $p \ge 3$, we can construct a special set of units in $Z[\zeta_p]$.

Proposition 4.24. *Let p be a prime number, $p \ge 3$. For each integer j, $1 \le j \le p-2$, the element $1 + \zeta_p + \zeta_p^2 + \cdots + \zeta_p^j$ is a unit in $Z[\zeta_p]$.*

Proof. For $1 \le j \le p-2$, $1 + \zeta_p + \zeta_p^2 + \cdots + \zeta_p^j$ has inverse

$$\frac{1-\zeta_p}{1-\zeta_p^{j+1}} = \frac{1-\zeta_p^{k(j+1)}}{1-\zeta_p^{j+1}} = 1 + \zeta_p^{j+1} + \zeta_p^{2(j+1)} + \cdots + \zeta_p^{(k-1)(j+1)},$$

with k chosen so that $k(j+1) \equiv 1 \bmod p$. □

The units constructed in Proposition 4.24 are the **circular units** in $Z[\zeta_p]$.

In general, the structure of the group of units of a ring of integers is given by a classical result of Dirichlet.

Proposition 4.25 (Dirichlet's Unit Theorem). *Let K be a finite extension of $\mathbb{Q}$ of degree n. Then*

$$U(R) \cong W \times Z^{r+s-1},$$

where r is the number of embeddings of K into $\mathbb{R}$, and s is the number of pairs of conjugate embeddings of K into $\mathbb{C}$ and W is the subgroup of $R^\times$ generated by the roots of unity in R.

Proof. For a proof see [Cassels and Fröhlich (1967), Chapter II, §18], or [Samuel (2008), §4.4]. □

The factor Z^{r+s-1} is a product of $r+s-1$ infinite cyclic groups

$$\langle \sigma_1 \rangle \times \langle \sigma_2 \rangle \times \cdots \times \langle \sigma_{r+s-1} \rangle.$$

The generators $\sigma_1, \sigma_2 \ldots, \sigma_{r+s-1}$ correspond to actual units in R which form a **set of fundamental units of** R, denoted as $f_1, f_2, \ldots, f_{s+r-1}$. Consequently every unit u in R can be written as

$$u = w f_1^{e_1} f_2^{e_2} \cdots f_{r+s-1}^{e_{r+s-1}}$$

for some $w \in W$ and integers $e_1, e_2, \ldots, e_{r+s-1}$.

Example 4.7. By Proposition 4.21(i) the ring of integers of $\mathbb{Q}(\sqrt{2})$ is $Z[\sqrt{2}]$. By Dirichlet's Unit Theorem, $U(Z[\sqrt{2}]) \cong \langle -1 \rangle \times Z$. More precisely, $f = 1 + \sqrt{2}$ is a fundamental unit of R and so

$$U(Z[\sqrt{2}]) = \{\pm(1+\sqrt{2})^n : n \in Z\}.$$

Example 4.8. By Proposition 4.22 the ring of integers of $\mathbb{Q}(\zeta_p)$ is $Z[\zeta_p]$. For $p \geq 3$, Dirichlet's Unit Theorem says that

$$U(Z[\zeta_p]) \cong W \times Z^{(p-3)/2},$$

with $W = \{\pm 1, \pm\zeta_p, \pm\zeta_p^2, \ldots, \pm\zeta_p^{p-1}\}$. A unit $u \in U(Z[\zeta_p])$ can be written in the form

$$u = w f_1^{e_1} f_2^{e_2} \cdots f_{(p-3)/2}^{e_{(p-3)/2}},$$

where $w \in W$ and $f_1, f_2, \ldots, f_{(p-3)/2}$ are fundamental units. For $p = 3$, there are no fundamental units and

$$U(Z[\zeta_3]) = W = \{\pm 1, \pm\zeta_3, \pm\zeta_3^2\}.$$

For $p = 5$,

$$U(Z[\zeta_5]) \cong W \times Z$$

with $W = \{\pm 1, \pm\zeta_5, \pm\zeta_5^2, \pm\zeta_5^3 \pm \zeta_5^4\}$ and fundamental unit $f_1 = 1 + \zeta_5$. The circular units $1 + \zeta_5$, $1 + \zeta_5 + \zeta_5^2$, and $1 + \zeta_5 + \zeta_5^2 + \zeta_5^3$ can be written

$$1 + \zeta_5 = f_1,$$

$$1 + \zeta_5 + \zeta_5^2 = -\zeta_5^3 f_1,$$

$$1 + \zeta_5 + \zeta_5^2 + \zeta_5^3 = -\zeta_5^4.$$

Since $Z[\zeta_3]$ has only trivial units, so does the group ring ZC_3. To prove this we need a lemma.

Lemma 4.5. *Let $C_p = \langle\sigma\rangle$. There is an isomorphism of rings $\phi : \mathbb{Q}C_p \to \mathbb{Q} \times \mathbb{Q}(\zeta_p)$ given by $\sigma \mapsto (1, \zeta_p)$.*

Proof. Let $p(x) = x^{p-1} + x^{p-2} + \cdots + x + 1$. In the polynomial ring $\mathbb{Q}[x]$, $\gcd(x - 1, p(x)) = 1$, and so, by Proposition 2.21,

$$(x - 1) + (p(x)) = \mathbb{Q}[x].$$

Now by the Chinese Remainder Theorem for Rings there is an isomorphism of rings

$$\mathbb{Q}[x]/(x^p - 1) \cong \mathbb{Q}[x]/(x - 1) \times \mathbb{Q}[x]/(p(x))$$

defined by $x+(x^p-1) \mapsto (x+(x-1), x+(p(x)))$. Since $\mathbb{Q}C_p \cong \mathbb{Q}[x]/(x^p-1)$, $\mathbb{Q} \cong \mathbb{Q}[x]/(x - 1)$ and $\mathbb{Q}(\zeta_p) \cong \mathbb{Q}[x]/(p(x))$, there is an isomorphism

$$\phi : \mathbb{Q}C_p \to \mathbb{Q} \times \mathbb{Q}(\zeta_p),$$

defined by $\phi(\sigma) = (1, \zeta_p)$. This isomorphism restricted to the ring ZC_p yields the ring injection (only):

$$ZC_p \to Z \times Z[\zeta_p],$$

given as $\sigma \mapsto (1, \zeta_p)$. □

Proposition 4.26. *Let $C_3 = \langle\sigma\rangle$. The group of units in ZC_3 is $\{\pm 1, \pm\sigma, \pm\sigma^2\}$.*

Proof. The isomorphism $\phi : \mathbb{Q}C_3 \to \mathbb{Q} \times \mathbb{Q}(\zeta_3)$, $\phi(\sigma) = (1, \zeta_3)$, of Lemma 4.5, restricted to the ring ZC_3 yields the ring injection

$$ZC_3 \to Z \times Z[\zeta_3],$$

given as $\sigma \mapsto (1, \zeta_3)$. Thus if $a+b\sigma+c\sigma^2$ is a unit of ZC_3, then $a+b+c = \pm 1$ and $a + b\zeta_3 + c\zeta_3^2 = \pm 1$, $\pm\zeta_3$, or $\pm\zeta_3^2$. Multiplying by -1, if necessary, we can assume without loss of generality that $a + b + c = 1$. Now,

$$(a + b\zeta_3 + c\zeta_3^2) - (a + b + c) = b(\zeta_3 - 1) + c(\zeta_3^2 - 1)$$

is divisible by $\zeta_3 - 1$, hence $a = \pm 1$, $b = c = 0$, $b = \pm 1$, $a = c = 0$, or $a = b = 0$, $c = \pm 1$. □

We ask: does the group ring ZC_p, $p \geq 5$ have non-trivial units?

Proposition 4.27. *For $p \geq 5$ the group ring ZC_p has non-trivial units.*

Proof. The isomorphism of Lemma 4.5 restricted to the ring ZC_p yields the ring injection

$$ZC_p \to Z \times Z[\zeta_p],$$

defined as $\sigma \mapsto (1, \zeta_p)$. Let j, $1 \leq j \leq p - 2$, be an integer and let

$$u_j = (1 + \sigma + \sigma^2 + \cdots + \sigma^j)^{p-1} - l(1 + \sigma + \sigma^2 + \cdots + \sigma^{p-1}),$$

where l is the integer that satisfies $(j + 1)^{p-1} = 1 + lp$. Then the image of u_j in $Z \times Z[\zeta_p]$ is

$$(1, (1 + \zeta_p + \zeta_p^2 + \cdots + \zeta_p^j)^{p-1}),$$

which is a unit in $Z \times Z[\zeta_p]$, since the second component is the $(p - 1)$th power of a circular unit in $Z[\zeta_p]$. It follows that u_j, $1 \leq j \leq p - 2$, is a non-trivial unit in ZC_p. □

Example 4.9. In ZC_5 one has that

$$\begin{aligned} u_1 &= (1 + \sigma)^4 - 3(1 + \sigma + \sigma^2 + \sigma^3 + \sigma^4) \\ &= -2 + \sigma + 3\sigma^2 + \sigma^3 - 2\sigma^4 \end{aligned}$$

is a non-trivial unit in ZC_5 with

$$u_1^{-1} = 1 + \sigma - 2\sigma^2 + 3\sigma^3 - 2\sigma^4.$$

4.4 The Noetherian Property of the Ring of Integers

In this section we show that the ring of integers R of a simple algebraic extension $K/\mathbb{Q}$ is a Noetherian ring by showing that R has the ACC.

* * *

Let $K = \mathbb{Q}(\alpha)$ be a simple algebraic extension of $\mathbb{Q}$ of degree l, let R denote the ring of integers of K, and let $\{b_1, b_2, \ldots, b_l\}$ be a $\mathbb{Q}$-basis for K. Let $\beta \in K$. There exist rationals $q_{i,j} \in \mathbb{Q}$, $1 \leq i, j \leq l$ for which

$$\begin{aligned}\beta b_1 &= q_{1,1}b_1 + q_{1,2}b_2 + \cdots + q_{1,l}b_l\\ \beta b_2 &= q_{2,1}b_1 + q_{2,2}b_2 + \cdots + q_{2,l}b_l\\ &\vdots\\ \beta b_l &= q_{l,1}b_1 + q_{l,2}b_2 + \cdots + q_{l,l}b_l.\end{aligned}$$

The **trace map** $\mathrm{tr}_{K/\mathbb{Q}} : K \to \mathbb{Q}$ is defined as $\mathrm{tr}_{K/\mathbb{Q}}(\beta) = \sum_{i=1}^{l} q_{i,i}$.

Lemma 4.6. *The trace map $tr_{K/\mathbb{Q}} : K \to \mathbb{Q}$ defined above does not depend on the choice of basis for K over $\mathbb{Q}$.*

Proof. Let $\beta \in K$, let $\{b_1, b_2, \ldots, b_l\}$ be a $\mathbb{Q}$-basis for K and let Q be the matrix in $\mathrm{Mat}_l(\mathbb{Q})$ for which

$$Q\begin{pmatrix} b_1\\ b_2\\ \vdots\\ b_l\end{pmatrix} = \begin{pmatrix} \beta b_1\\ \beta b_2\\ \vdots\\ \beta b_l\end{pmatrix}.$$

Then $\mathrm{tr}_{K/\mathbb{Q}}(\beta) = \mathrm{tr}(Q)$, the ordinary trace of the matrix Q. Now let $\{b'_1, b'_2, \ldots, b'_l\}$ be some other basis for K over $\mathbb{Q}$. Then there exists an invertible matrix $U \in \mathrm{Mat}_l(\mathbb{Q})$ with inverse U^{-1} so that

$$U\begin{pmatrix} b_1\\ b_2\\ \vdots\\ b_l\end{pmatrix} = \begin{pmatrix} b'_1\\ b'_2\\ \vdots\\ b'_l\end{pmatrix}, \quad U^{-1}\begin{pmatrix} b'_1\\ b'_2\\ \vdots\\ b'_l\end{pmatrix} = \begin{pmatrix} b_1\\ b_2\\ \vdots\\ b_l\end{pmatrix}.$$

One has

$$UQU^{-1}\begin{pmatrix} b'_1\\ b'_2\\ \vdots\\ b'_l\end{pmatrix} = \begin{pmatrix} \beta b'_1\\ \beta b'_2\\ \vdots\\ \beta b'_l\end{pmatrix},$$

and so,

$$\mathrm{tr}_{K/\mathbb{Q}}(\beta) = \mathrm{tr}(UQU^{-1}) = \mathrm{tr}(Q) = \mathrm{tr}_{K/\mathbb{Q}}(\beta)$$

since the trace of a matrix is invariant under conjugation. □

Proposition 4.28. *The trace map defined as above coincides with the trace map given in Definition 4.2.*

Proof. Let $K = \mathbb{Q}(\alpha)$ and let $g_1, g_2, \ldots, g_l$ denote the distinct set of embeddings $g_i : K \to \mathbb{C}$. Let $\beta \in K$, let $s = [\mathbb{Q}(\beta) : \mathbb{Q}]$. Let $p(x)$ be the irreducible polynomial of β over $\mathbb{Q}$ with distinct roots $\beta = \beta_1, \beta_2, \ldots \beta_s$ and corresponding embeddings $\tau_j : \mathbb{Q}(\beta) \to \mathbb{C}$, $1 \le j \le s$. Let $t = [\mathbb{Q}(\alpha) : \mathbb{Q}(\beta)]$ and let $q(x)$ be the irreducible polynomial for α over $\mathbb{Q}(\beta)$ with distinct roots $\alpha = \alpha_1, \alpha_2, \ldots, \alpha_t$.

By Lemma 4.6, the trace map does not depend on the choice of basis, so choose the basis $\{\beta^j \alpha^i\}$, $0 \le j \le s-1$, $0 \le i \le t-1$, for K over $\mathbb{Q}$. Using this basis

$$\mathrm{tr}_{K/\mathbb{Q}}(\beta) = t \cdot \sum_{j=1}^{s} \tau_j(\beta) = \sum_{i=1}^{l} g_i(\beta),$$

which is precisely the trace map given in Definition 4.2. □

Proposition 4.29. *The trace map defines a symmetric non-degenerate bilinear form on K:*

$$B : K \times K \to \mathbb{Q}, \quad B(x, y) = tr(xy),$$

for $x, y \in K$.

Proof. This is straightforward if one takes $\mathrm{tr}_{K/\mathbb{Q}} : K \to \mathbb{Q}$ defined as $\mathrm{tr}_{K/\mathbb{Q}} = \sum_{i=1}^{l} g_i$. □

Proposition 4.30. *Let $K = \mathbb{Q}(\alpha)$ be a finite extension of $\mathbb{Q}$ with ring of integers R. Then the bilinear form given above restricts to a bilinear form $B : R \times R \to Z$.*

Proof. Take $\mathrm{Tr}_{K/\mathbb{Q}} = \sum_{i=1}^{l} g_i$. Let $\gamma, \beta \in R$. Then $\gamma\beta \in R$. Now, by Proposition 4.29, $\mathrm{Tr}_{K/\mathbb{Q}}(\gamma\beta) = B(\gamma, \beta) \in Z$. □

Now let J be a non-zero ideal of R and let $\mathcal{S}$ be the collection of all bases $\mathcal{B}$ for $K/\mathbb{Q}$ which are contained in J. By Lemma 4.1, $\mathcal{S}$ is nonempty. For each basis $\mathcal{B} = \{b_1, b_2, \ldots, b_l\}$ in $\mathcal{S}$, let

$$N_{\mathcal{B}} = Zb_1 \oplus Zb_2 \oplus \cdots \oplus Zb_l \subseteq J$$

be the free Z-module with basis $\mathcal{B}$. We compute $\text{disc}(N_{\mathcal{B}})$ with respect to the bilinear form of Proposition 4.29 using the basis $\{b_1, b_2, \ldots, b_l\}$. By Proposition 4.30, $\text{disc}(N_{\mathcal{B}})$ is generated by a non-zero integer which we identify with $\text{disc}(N_{\mathcal{B}})$. The collection

$$\{|\text{disc}(N_{\mathcal{B}})|\}_{\mathcal{B}\in\mathcal{S}},$$

is a non-empty set of positive integers, and as such, has a smallest element. Let $\mathcal{M} = \{m_1, m_2, \ldots, m_l\}$ denote a basis in $\mathcal{S}$ which corresponds to the smallest integer $|\text{disc}(N_{\mathcal{M}})|$.

Lemma 4.7. $J = N_{\mathcal{M}}$.

Proof. We only need to show that $J \subseteq N_{\mathcal{M}}$. Let $a \in J$. Since $\{m_i\}$ is a basis for $K/\mathbb{Q}$,

$$a = q_1m_1 + q_2m_2 + \cdots + q_lm_l,$$

for elements $q_i \in \mathbb{Q}$. We claim that each q_i is an integer. By way of contradiction, let's assume that $q_j \notin Z$ for some j. Without loss of generality, we can assume that $j = 1$. Note that $q_1 = \eta + \iota$ for some $\eta \in Z$ and ι with $0 < \iota < 1$. Set $m_1' = a - \eta m_1$ and $m_i' = m_i$ for $2 \leq i \leq l$. Then $\mathcal{M}' = \{m_i'\}$ is a basis for $K/\mathbb{Q}$ which is contained in J.

Let $N_{\mathcal{M}'} = Zm_1' \oplus Zm_2' \oplus \cdots \oplus Zm_l'$. Then $N_{\mathcal{M}'} \subseteq N_{\mathcal{M}}$. The matrix which multiplies the basis $\{m_i\}$ to give the basis $\{m_i'\}$ is

$$\begin{pmatrix} \iota & q_2 & q_3 & \cdots & q_l \\ 0 & 1 & 0 & & 0 \\ 0 & 0 & 1 & & 0 \\ \vdots & & & & \vdots \\ 0 & 0 & \cdots & 0 & 1 \end{pmatrix},$$

thus the module index is

$$[N_{\mathcal{M}} : N_{\mathcal{M}'}] = \iota Z.$$

Now by Proposition 3.34

$$\text{disc}(N_{\mathcal{M}'}) = (\iota^2 Z)\text{disc}(N_{\mathcal{M}}),$$

which contradicts the minimality of $|\text{disc}(N_{\mathcal{M}})|$ since $\iota^2 < 1$. Thus each q_i is an integer, and so $J = N_{\mathcal{M}}$. □

The following proposition will imply that R is a Noetherian ring.

Proposition 4.31. *For any non-zero ideal J of R, R/J is a finite ring.*

Proof. Let $a \in J \cap Z^+$. (Why is $J \cap Z^+$ non-empty?) Since the ring homomorphism $R/(a) \to R/J$ is surjective, we only need to show that $R/(a)$ is finite. By Lemma 4.7,

$$J = Zm_1 \oplus Zm_2 \oplus \cdots \oplus Zm_l,$$

for some elements $m_i \in J$. Now,

$$S = \{a_1 m_1 + a_2 m_2 + \cdots + a_l m_l : \ 0 \le a_i \le a\}$$

is a set of coset representatives for $R/(a)$. Note that $|S| = a^l < \infty$, so that $|R/(a)| = a^l$. □

Proposition 4.32. *Let R be the ring of integers of $K = \mathbb{Q}(\alpha)$. Then R is Noetherian.*

Proof. We show that R has the ACC. Let $I_1 \subseteq I_2 \subseteq I_3 \subseteq \cdots$ be an ascending chain of ideals of R. Suppose there is no integer $m \ge 0$ for which $I_m = I_{m+1} = I_{m+2} = \cdots$. Then I_n is non-trivial for some n and the quotient ring R/I_n has an infinite number of ideals $I_j + I_n$ for $j \ge n$. This is impossible by Proposition 4.31. □

4.5 Dedekind Domains

In §4.4 we showed that the ring of integers R is Noetherian. In this section we show that R has additional properties: it is a Dedekind domain. We show that if R is any Dedekind domain, then the localization of R_P, P a prime ideal of R, is a PID. From this result we construct the class group of R which is the group of fractional ideals over R modulo the principal fractional ideals over R. We include some examples of class groups of rings of integers.

* * *

Let R be an integral domain with $K = \text{Frac}(R)$. An element $\alpha \in K$ is **integral** over R if it is a zero of a monic polynomial with coefficients in R (cf. §4.3). The set of all elements of K that are integral over R is the **integral closure of** R **in** K. The integral domain R is **integrally closed** if the integral closure of R in K is R.

Proposition 4.33. *Let R be the ring of integers of a simple algebraic extension K of $\mathbb{Q}$. Then R is an integrally closed Noetherian ring in which each non-zero prime ideal is maximal.*

Proof. We already know that R is Noetherian (Proposition 4.32). By Proposition 4.19, R is an integral domain with $\text{Frac}(R) = K$. We need to show that R is integrally closed. To this end, let $\alpha \in K$ be a root of the monic polynomial of degree t over R:

$$f(x) = a_0 + a_1x + a_2x^2 + \cdots + a_{t-1}x^{t-1} + x^t.$$

Now, there is an inclusion of rings $R[\alpha] \subseteq K$ with $R[\alpha]$ finitely generated as an R-module (a generating set is $\{1, \alpha, \alpha^2, \ldots, \alpha^{t-1}\}$). Moreover, $\alpha R[\alpha] \subseteq R[\alpha]$. Let $\{b_1, b_2, \ldots, b_n\}$ be a Z-basis for R (which exists by Lemma 4.7). Now, α is an element of the finitely generated Z-module $M = Z[b_1, b_2, \cdots, b_n][\alpha]$. One has an inclusion of rings $M \subseteq K$ with $\alpha M \subseteq M$ and so by Proposition 4.15, α is integral over Z. It follows that $\alpha \in R$.

Finally, let J be a non-zero prime ideal. Then by Proposition 4.31, R/J is a finite integral domain, hence a field by Proposition 2.3. It follows that J is maximal. □

An integral domain that is an integrally closed Noetherian ring in which each non-zero prime ideal is maximal is a **Dedekind domain**. The ring of integers R is a Dedekind domain but so is the localization R_P.

Proposition 4.34. *Let R be the ring of integers of a finite extension K of $\mathbb{Q}$. Let P be a prime ideal of R and let R_P be the localization of R at P. Then R_P is a Dedekind domain.*

Proof. We first show that R_P is Noetherian. Let I be an ideal of R_P, then $I = (I \cap R)R_P$, and hence $I = JR_P$ for some ideal J of R. Consequently, I is finitely generated over R_P since R is Noetherian. Next, let $\alpha \in K$ be integral over R_P. Then α is a root of the polynomial

$$f(x) = r_0s^{-1} + r_1s^{-1}x + r_2s^{-1}x^2 + \cdots + r_{t-1}s^{-1}x^{t-1} + x^t,$$

with $r_i \in R$, $s \in R \backslash P$. Now $g(s\alpha) = 0$ where $g(x) = s^t f(x) \in R[x]$. Thus $s\alpha$ is integral over R. Since R is integrally closed, $s\alpha \in R$, whence $\alpha \in R_P$ and so R_P is integrally closed.

Finally, let I be a non-zero prime ideal of R_P. Then $I \subseteq PR_P$ since PR_P is the unique maximal ideal in R_P. Let J be the non-zero ideal of R with $I = JR_P$. Then $J \subseteq P$. But since J is a non-zero prime ideal of R, J is maximal and so $J = P$. It follows that $I = PR_P$ and so I is a maximal ideal of R_P. □

Quite generally, let R be an integral domain with field of fractions K. Then K is an R-module. A **fractional ideal** over R is a non-zero R-submodule J of K of the form $J = cI$ where $c \in K^\times$ and I is an ideal of R. A **principal fractional ideal** is a fractional ideal over R of the form cR for some $c \in K^\times$.

For example, $\frac{1}{2}Z = \{n/2 : \ n \in Z\}$ is a fractional ideal over Z, in fact it is principal. Certainly, any ideal I of R is a fractional ideal over R. If I, J are fractional ideals, then $I + J$ and IJ are fractional ideals.

Lemma 4.8. *Let R be a Dedekind domain with $K =$ Frac(R). A non-zero submodule J of K is a fractional ideal if and only if it is finitely generated as an R-module.*

Proof. Let J be a non-zero submodule of K of the form cI, $c \in K^\times$. Then J is finitely generated since I is finitely generated (R is Noetherian). Conversely, suppose J is a non-zero submodule of K which is finitely generated as an R-module. Write $J = Rq_1 + Rq_2 + \cdots + Rq_l$ for $q_1, q_2, \ldots, q_l \in K$. There is a generating set for J of the form $\{a_1/q, a_2/q, \ldots, a_l/q\}$ for some $a_i \in R$, $q \in R\backslash\{0\}$. Consequently, $J = cI$ where $c = q^{-1}$ and $I = (a_1, a_2, \ldots, a_l)$. □

Let R be a Dedekind domain with $K = \text{Frac}(R)$ and let J be a fractional ideal over R. Define

$$J^{-1} = \{x \in K : \ xJ \subseteq R\},$$

$$\tilde{J} = \{x \in K : xJ \subseteq J\}.$$

Lemma 4.9. *J^{-1} and $\tilde{J}$ are fractional ideals over R.*

Proof. Let y be a non-zero element of J. Then $J^{-1}y \subseteq R$, hence $J^{-1} \subseteq y^{-1}R$. Now, J^{-1} is a submodule of the finitely generated R-module $y^{-1}R$.

Thus J^{-1} is finitely generated by Corollary 3.1, and is a fractional ideal by Lemma 4.8. A similar argument shows that $\tilde{J}$ is fractional. □

Here is one reason why Dedekind domains are so important.

Proposition 4.35. *Let R be a Dedekind domain. Let P be a non-zero prime ideal of R and let R_P denote the localization of R at P. Then the unique maximal ideal $m = PR_P$ of R_P is a principal ideal.*

Proof. *Step 1.* First note that m is a fractional ideal over R_P and so, $\tilde{m} = \{x \in K : \ xm \subseteq m\}$ is a fractional ideal over R_P by Lemma 4.9. We claim that $\tilde{m} = R_P$. Certainly $R_P \subseteq \tilde{m}$, so it is a matter of checking the reverse inclusion. Let $x \in \tilde{m}$. Then $R_P[x]$ is an R_P-submodule of $\tilde{m}$. Since R_P is a Noetherian ring (Proposition 4.34) $R_P[x]$ is finitely generated as an R_P-module by Corollary 3.1. Now $xR_P[x] \subseteq R_P[x]$, and so an application of Proposition 4.15 shows that x is integral over R_P. Since R_P is integrally closed, $x \in R_P$.

Step 2. We show that $m^{-1} \neq R_P$, where $m^{-1} = \{x \in K : \ xm \subseteq R_P\}$. To this end, let $\mathcal{J}$ be the collection of all non-zero ideals J of R_P that satisfy

$$J^{-1} \neq R_P.$$

Then $\mathcal{J}$ is non-empty set since $(r) \in \mathcal{J}$ for any non-zero $r \in m$. Suppose that $\mathcal{J}$ has no maximal element. Then (r) is not maximal, and so there is an element J_1 in $\mathcal{J}$ with $(r) \subset J_1$. But J_1 is not maximal, and so there is an element $J_2 \in \mathcal{J}$ with $(r) \subset J_1 \subset J_2$. Continuing in this manner, we create an ascending sequence of ideals of R_P that does not stop, a contradiction of the ACC of the Noetherian ring R_P. So, $\mathcal{J}$ has a maximal element which we denote as M.

We claim that $M = m$. To this end, we show that M is a prime ideal of R_P. Let $x, y \in R_P$ with $xy \in M$. Assume that $x \notin M$. Since $M \in \mathcal{J}$, there exists an element $z \in K$ with $z \in M^{-1}$, $z \notin R_P$. We claim that $yz \in R_P$. Now, $yz(xR_P + M) \subseteq R$ and so, if $zy \neq R_P$, then $xR_P + M \in \mathcal{J}$ with $M = xR_P + M$. Consequently, $x \in M$, a contradiction. Now $yz \in R_P$, and so,

$$z(yR_P + M) = yzR_P + zM \subseteq R_P$$

with $z \notin R_P$. Thus the fractional ideal $yR_P + M$ satisfies the requirements for membership in $\mathcal{J}$. Thus $M = yR_P + M$, and so $y \in M$.

So M is prime and hence maximal by Proposition 4.34. Thus $m = M \in \mathcal{J}$ and so, $m^{-1} \neq R_P$.

Step 3. We show that m is principal. Note that

$$m \subseteq m^{-1}m \subseteq R_P,$$

and since m is maximal, either $m = m^{-1}m$ or $m^{-1}m = R_P$. In the former case, $m^{-1} \subseteq \tilde{m}$ and thus $m^{-1} = \tilde{m}$ since $\tilde{m} \subseteq m^{-1}$. But $m^{-1} = R_P$ by Step 1, contradicting Step 2. And so, we conclude that

$$m^{-1}m = R_P.$$

Now consider the ideal

$$N = \bigcap_{n=1}^{\infty} m^n.$$

If $N \neq 0$, then N is a fractional ideal over R_P and hence, $\tilde{N} = \{x \in K : xN \subseteq N\}$, is a fractional ideal over R_P by Lemma 4.9. Now the method of Step 1 shows that $\tilde{N} = R_P$. Observe that $m^{-1} \subseteq \tilde{N}$. Thus, $R_P \subseteq m^{-1} \subseteq R_P$, and so, $m^{-1} = R_P$, a contradiction. Thus $N = \{0\}$ which says that there exists an element $\pi \in m$ for which $\pi R_P \subseteq m$ and $\pi R_P \not\subseteq m^2$. Thus $\pi m^{-1} \subseteq R_P$; πm^{-1} is an ideal of R_P with $\pi m^{-1} \not\subseteq m$. Thus $\pi m^{-1} = R_P$ and so, $m = \pi R_P$. □

Proposition 4.36. *Let R be a Dedekind domain and let P be a prime ideal of R. Then R_P is a PID.*

Proof. Let $PR_P = \pi R_P$ and let I be a non-zero ideal of R_P. For $n \geq 0$, $I\pi^{-n}$ is a finitely generated R_P-module with $I\pi^{-n} \subseteq I\pi^{-n-1}$. Suppose that $I\pi^{-n} = I\pi^{-n-1}$ for some n. Then $\pi^{-1}(I\pi^{-n}) = I\pi^{-n}$, and so π^{-1} is integral over R_P by Proposition 4.15. Hence $\pi^{-1} \in R_P$ since R_P is integrally closed. Now $1 = \pi\pi^{-1} \in \pi R_P$, which is impossible. Consequently, there is a strictly increasing sequence

$$I \subset I\pi^{-1} \subset I\pi^{-2} \subset \cdots$$

Since R_P is Noetherian, there exists an integer $n \geq 0$ for which $I\pi^{-n}$ is an ideal of R_P and $I\pi^{-n-1} \not\subseteq R_P$. Hence $I\pi^{-n} \not\subseteq \pi R_P$, thus, $I\pi^{-n} = R_P$ by Corollary 2.3. It follows that $I = \pi^n R_P$; I is principal. □

Proposition 4.37. *Let R be a Dedekind domain and let J be a fractional ideal over R. Then $J^{-1}J = R$.*

Proof. Let $J_P = JR_P$, $J_P^{-1} = J^{-1}R_P$. Then J_P and J_P^{-1} are fractional ideals over R_P. Since R_P is a PID, there exists $c \in K^\times$ for which $J_P = cR_P$ and $J_P^{-1} = c^{-1}R_P$. Thus

$$J_P J_P^{-1} = cR_P c^{-1} R_P = R_P.$$

Consequently,

$$JJ^{-1} = \bigcap_P J_P J_P^{-1} = \bigcap_P R_P = R,$$

by Proposition 3.20. □

Proposition 4.38. *Let R be a Dedekind domain and let $\mathcal{F}(R)$ denote the collection of fractional ideals over R. Then $\mathcal{F}(R)$ is an abelian group under the binary operation $\star$ defined as $I \star J = IJ$, for $I, J \in \mathcal{F}(R)$.*

Proof. Clearly $\star$ is associative and commutative. For an identity element, take $e = R$. Lastly, by Proposition 4.37 we can take J^{-1} to be the inverse of J under $\star$. □

Proposition 4.39. *Let R be a Dedekind domain. Let $\mathcal{PF}(R)$ denote the collection of principal fractional ideals over R. Then $\mathcal{PF}(R) \leq \mathcal{F}(P)$.*

Proof. Exercise. □

The quotient group $\mathcal{F}(R)/\mathcal{P}F(R)$ is the **class group of** R, denoted as $\mathcal{C}(R)$. Given an fractional ideal J over R in K, the left coset $J\mathcal{P}F(R) \in \mathcal{C}(R)$ is the **ideal class** of J. The class group $\mathcal{C}(R)$ is a finite abelian group [Samuel (2008), §4.3]; the order of $\mathcal{C}(R)$, denoted as h_R, is the **class number** of R.

The significance of the class group $\mathcal{C}(R)$ is immediate: a Dedekind domain R is a PID if and only if $h_R = 1$. In this sense, the value of h_R measures the degree to which R fails to be a PID.

Example 4.10. Let m be a square-free integer, $m < 0$ and let $K = \mathbb{Q}(\sqrt{m})$. Then $h_R = 1$ if and only if

$$m = -1, -2, -3, -7, -11, -19, -43, -67, -163.$$

This result is due to [Stark (1967)]. In the case that $m = -5$, $R = Z[\sqrt{-5}]$ is not a PID and we have $h_R = 2$. In the case that $m = -23$, $R = Z[(1 + \sqrt{-23})/2]$ is not a PID, as we have seen. In fact, $h_R = 3$.

Example 4.11. Let m be a square-free integer $2 \leq m \leq 50$. Then $h_R = 1$ for the following values of m [Sloane (2015), A003172]:

$$m = 2, 3, 5, 6, 7, 11, 13, 14, 17, 19, 21, 22, 23, 29, 31, 33, 37, 38, 41, 43, 46, 47.$$

In the case that $m = 10$, one has $h_R = 2$ [Sloane (2015), A094619].

Example 4.12. Let $p \geq 2$ be a prime number and let $K = \mathbb{Q}(\zeta_p)$. Then as we have shown, $R = Z[\zeta_p]$. The complete list of primes for which $h_R = 1$ is:

$$p = 2, 3, 5, 7, 11, 13, 17, 19$$

[Washington (1997), Theorem 11.1].

4.6 Unique Factorization of Ideals

In this section we show that each non-zero proper ideal of a Dedekind domain can be factored into a product of prime ideals in a unique way; this factorization generalizes the familiar notion that each positive integer $m \in Z$ factors uniquely into a product primes $m = p_1^{e_1} p_2^{e_2} \cdots p_k^{e_k}$. We introduce ramified and unramified primes and the Hilbert class field.

Proposition 4.40. *Let R be a Dedekind domain, and let J be a non-trivial proper ideal of R. Then J can be factored into a product of prime ideals of R*

$$J = P_1^{e_1} P_2^{e_2} \cdots P_k^{e_k},$$

where e_i are positive integers. Moreover, this factorization is unique up to a re-ordering of the factors $P_i^{e_i}$.

Proof. In the first part of the proof we find a prime ideal P_1 and an ideal J_1 for which $J = J_1 P_1$ with $J \subset J_1$.

Since R is Noetherian, §2.8, Exercise 46 applies to show that J is contained in a prime ideal P_1 of R. P_1 is a fractional ideal of R with $P_1^{-1} P_1 = R$, thus there exists elements $q_i \in P_1^{-1}$ and $r_i \in P_1$ for which $1 = \sum_{i=1}^{n} q_i r_i$. Now $J \subseteq P_1$ implies $P_1^{-1} J \subseteq R$, thus $P_1^{-1} J$ is an ideal of R, with $J = JR = J(P_1^{-1} P_1) = (P_1^{-1} J) P_1$.

Set $J_1 = P_1^{-1} J$. Let $a \in J$, then $a = \sum_{i=1}^{l} a q_i r_i$, and so, $J \subseteq J_1$. Now, if $J = J_1$, then $P_1 J = P_1 J_1 = P_1 P_1^{-1} J = J$. Multiplying on the right by J^{-1} yields $P_1 = R$. Thus $J \subset J_1$.

If $J_1 = R$, then we have proved the proposition. If J_1 is proper, then since R is Noetherian, J_1 is contained in a prime ideal P_2, and as above, $J_1 = J_2 P_2$, for some J_2 with $J_1 \subset J_2$. Now, $J = J_2 P_1 P_2$. Again, we are done if $J_2 = R$. Else, we repeat the process obtaining a collection of ideals $J, J_1, J_2 \ldots$

We claim there exists an integer k for which $J_k = R$ and consequently, the process stops with $J = P_1P_2 \cdots P_k$. Suppose no such k exists. Then there is an ascending chain of ideals

$$J \subset J_1 \subset J_2 \subset J_3 \subset \cdots .$$

Each containment is proper and the chain does not stop; the ACC fails. This is impossible since R is Noetherian.

We leave the statement regarding uniqueness of the factorization as an exercise. □

Example 4.13. Let $K = \mathbb{Q}(\sqrt{-5})$. By Proposition 4.21, the ring of integers of K is $R = Z[\sqrt{-5}]$. As one can check, the ideal (2) in R factors uniquely into prime ideals as

$$(2) = (2, 1 + \sqrt{-5})^2,$$

and the ideal (3) factors uniquely into prime ideals as

$$(3) = (3, 1 + \sqrt{-5})(3, 1 - \sqrt{-5}).$$

Moreover, the ideals $(1 + \sqrt{-5})$ and $(1 - \sqrt{-5})$ factor uniquely as

$$(1 + \sqrt{-5}) = (2, 1 + \sqrt{-5})(3, 1 + \sqrt{-5}),$$

$$(1 - \sqrt{-5}) = (2, 1 - \sqrt{-5})(3, 1 - \sqrt{-5}).$$

Observe that the non-zero non-unit element $6 \in R$ factors into irreducible elements of R in two different ways:

$$6 = 2 \cdot 3,$$

$$6 = (1 + \sqrt{-5})(1 - \sqrt{-5}),$$

and so R is not a UFD (thus R is not a PID). The principal ideal (6) however, does factor uniquely into prime ideals,

$$(6) = (2, 1 + \sqrt{-5})(2, 1 - \sqrt{-5})(3, 1 + \sqrt{-5})(3, 1 - \sqrt{-5}).$$

Example 4.14. Let $K = \mathbb{Q}(\zeta_{p^n})$. By Proposition 4.23, the ring of integers of K is $R = Z[\zeta_{p^n}]$. From the factorization

$$\begin{aligned} x^{p^n} - 1 &= (x^{p^{n-1}})^p - 1 \\ &= (x^{p^{n-1}} - 1)((x^{p^{n-1}})^{p-1} + (x^{p^{n-1}})^{p-2} + \cdots + x^{p^{n-1}} + 1) \end{aligned}$$

we obtain the formula

$$p = \prod_{\substack{1 \le i \le p^n - 1, \\ \gcd(i,p)=1}} (1 - \zeta_{p^n}^i).$$

In fact, for all i, $1 \le i \le p^n - 1$, $\gcd(i,p) = 1$, the principal ideal $(1 - \zeta_{p^n}^i)$ equals the prinicipal ideal $(1 - \zeta_{p^n})$ which is prime. Thus the ideal (p) in R has unique factorization

$$(p) = (1 - \zeta_{p^n})^{p^{n-1}(p-1)}.$$

Let K be a finite extension of $\mathbb{Q}$ with ring of integers R; let L be a finite extension of K with ring of integers S. Let Q be a prime ideal of R. The ideal SQ of S factors uniquely into prime ideals of S,

$$SQ = P_1^{e_1} P_2^{e_2} \cdots P_m^{e_m}.$$

If $e_i > 1$ for some i, $1 \le i \le m$, then Q is a **ramified prime** of R. If $e_i = 1$ for all $1 \le i \le m$, then Q is an **unramified prime** of R. If every prime Q of R is unramified, then L is an **unramified extension** of K.

Proposition 4.41. *Let K be a finite extension of $\mathbb{Q}$, $[K : \mathbb{Q}] > 1$, with ring of integers R. Then there exists a prime number $p \in Z$ for which pZ is a ramified prime of Z.*

Proof. The proof is beyond the scope of this book. The interested reader should consult [Neukirch (1999), Theorem III.2.17]. □

Consequently, $\mathbb{Q}$ is the only unramified extension of $\mathbb{Q}$.

Let K be a finite extension of $\mathbb{Q}$. The **Hilbert Class Field** L of K is the maximal abelian unramified extension of K. The Hilbert Class Field L of K is a Galois extension with Galois group isomorphic to the class group $\mathcal{C}(R)$ of K, see [Ireland and Rosen (1990), Notes, p. 184].

Example 4.15. By Proposition 4.41, the Hilbert Class Field of $\mathbb{Q}$ is $\mathbb{Q}$.

Example 4.16. Let $K = \mathbb{Q}(\sqrt{-5})$. Then $R = \mathbb{Z}[\sqrt{-5}]$ with class group $\mathcal{C}(R) \cong C_2$. The Hilbert Class Field of K is $L = K(\mathrm{i})$ with ring of integers $S = R[(\mathrm{i} + \sqrt{-5})/2]$; L is a Galois extension of K with Galois group C_2.

The notions of UFD and PID are equivalent in Dedekind domains.

Proposition 4.42. *Let R be a Dedekind domain. Then R is a PID if and only if R is a UFD.*

Proof. Suppose that R is a PID. Then by Proposition 2.30, R is a UFD. To prove the converse, assume that R is UFD. Let J be a non-trivial proper ideal of R. We claim that J is principal. By Proposition 4.40 there exists a unique factorization

$$J = P_1^{e_1} P_2^{e_2} \cdots P_k^{e_k},$$

where $P_1, P_2, \ldots, P_k$ are prime ideals of R. So we can show that J is principal by proving that every prime ideal of R is principal. To this end, let P be a non-zero prime ideal of R. Let r be a non-zero element of P. Then if $(r) = P$ we are done. Else, assume that $(r) \subset P$. From the proof of Proposition 4.40, we conclude that there exists a non-trivial proper ideal I for which $(r) = IP$. Let

$$I = Q_1^{e_1} Q_2^{e_2} \cdots Q_l^{e_l}$$

be the factorization of I into prime ideals Q_i. Now

$$(r) = P Q_1^{e_1} Q_2^{e_2} \cdots Q_l^{e_l}.$$

But since R is a UFD, $r = q_1 q_2 \cdots q_m$ for irreducible elements $q_i \in R$. By Proposition 2.24 each ideal (q_i) is prime. Thus $(r) = (q_1)(q_2)\cdots(q_m)$ is a prime factorization. By uniqueness of prime factorizations, $P = (q_i)$ for some i. Thus P is principal which proves the proposition. □

4.7 Extensions of $\mathbb{Q}_p$

In this final section of the chapter we construct field extensions of the p-adic rationals $\mathbb{Q}_p$. Beginning with a simple algebraic extension $K/\mathbb{Q}$ with ring of integers R, we use the unique factorization of (p) in R to extend the p-adic absolute value $|\ |_p$ to an absolute value $|\ |_P$ on K, where P is a prime in the factorization of (p). Analogous to §2.7, we complete K with respect to the extension $|\ |_P$, resulting in a finite extension of fields $K_P/\mathbb{Q}_p$.

* * *

Let p be a prime number, let K be a simple algebraic extension of $\mathbb{Q}$ of degree $n = [K : \mathbb{Q}]$. Let R be the ring of integers of K. By Proposition 4.40 (p) factors uniquely into a product of prime ideals of R,

$$(p) = P_1^{e_1} P_2^{e_2} \cdots P_g^{e_g}. \tag{4.1}$$

For each i, $1 \le i \le g$, we define an absolute value $|\ |_{P_i}$ on K as follows. For $x = 0$, let $|x|_{P_i} = 0$. For $x = r/s \in K$, $r, s \in R$, $r \neq 0$, $s \neq 0$, let t_r be the integer $t_r \geq 0$ for which $(r) \subseteq P_i^{t_r}$, $(r) \not\subseteq P_i^{t_r+1}$; let t_s be the integer $t_s \geq 0$ for which $(s) \subseteq P_i^{t_s}$, $(s) \not\subseteq P_i^{t_s+1}$, and define

$$|x|_{P_i} = \frac{1}{p^{(t_r - t_s)/e_i}}.$$

Then as one can verify $|\ |_{P_i}$ is an absolute value on K which extends $|\ |_p$. Indeed, $|p|_{P_i} = \frac{1}{p}$ as required.

As we did with $\mathbb{Q}$ endowed with $|\ |_p$, we ask: does every $|\ |_{P_i}$-Cauchy sequence in K converge to a limit in K? In other words is K complete with respect to the absolute value $|\ |_{P_i}$? Not surprisingly, the answer is "no".

Proposition 4.43. *Let $K/\mathbb{Q}$ be a finite extension of degree n and suppose that $|\ |_{P_i}$ is an extension of the p-adic absolute value. The K is not complete with respect to $|\ |_{P_i}$.*

Proof. We construct a $|\ |_{P_i}$-Cauchy sequence in K that does not converge to a limit in K. There exists a prime $q \neq p$ for which $q > n$ and integers x_1 and a so that $x_1^q \equiv a \bmod p$, $a \not\equiv 0 \bmod p$, a is not a qth power in $\mathbb{Q}$. Then $p(x) = x^q - a$ is irreducible over $\mathbb{Q}$. Using the method of Proposition 2.49 we construct a $|\ |_{P_i}$-Cauchy sequence $\{x_1, x_2, x_3, \dots\}$ in K that converges to a limiting value α that is a root of $p(x)$. Now if $\alpha \in K$, then $[K : \mathbb{Q}] \geq q > n$, a contradiction. We conclude that K is not complete with respect to $|\ |_{P_i}$. $\square$

We form the completion of K with respect to $|\ |_{P_i}$ using the recipe of §2.7: First, we construct a field extension of K defined as

$$K_{P_i} = \mathcal{C}_{P_i}/\mathcal{N}_{P_i},$$

where $\mathcal{C}_{P_i}$ is the set of all $|\ |_{P_i}$-Cauchy sequences $\{x_n\}$ in K and

$$\mathcal{N}_{P_i} = \{\{x_n\} \in \mathcal{C}_{P_i} : \lim_{n\to\infty} x_n = 0\}.$$

We then show that the absolute value $|\ |_{P_i}$ extends uniquely to an absolute value on K_{P_i} which we also denote as $|\ |_{P_i}$. Next, we show that K is dense in K_{P_i}, and finally, we show that K_{P_i} is complete with respect to $|\ |_{P_i}$. The field K_{P_i} is the **completion of K with respect to $|\ |_{P_i}$.**

Proposition 4.44. *Let K be a simple algebraic extension of $\mathbb{Q}$ of degree $n = [K : \mathbb{Q}]$. Let p be a prime number and let $(p) = P_1^{e_1} P_2^{e_2} \cdots P_g^{e_g}$ be the prime factorization of (p). For each i, $1 \leq i \leq g$, the completion K_{P_i} is a finite extension of $\mathbb{Q}_p$ of degree n_i. We have the formula*

$$n = \sum_{i=1}^{g} n_i.$$

Proof. Write $K = \mathbb{Q}(\alpha)$ for some $\alpha \in K$ and let $p(x)$ be the monic irreducible polynomial for α. Let $p(x) = \prod_{j=1}^{l} p_j(x)$ be the factorization of $p(x)$ into monic irreducible polynomials over $\mathbb{Q}_p$. Let $n_j = \deg(p_j(x))$. We have

$$\begin{aligned} K \otimes_{\mathbb{Q}} \mathbb{Q}_p &\cong \mathbb{Q}[x]/(p(x)) \otimes_{\mathbb{Q}} \mathbb{Q}_p \\ &\cong \mathbb{Q}_p[x]/(p(x)) \\ &\cong \prod_{j=1}^{l} \mathbb{Q}_p[x]/(p_j(x)) \quad \text{by Proposition 2.19} \\ &= \prod_{j=1}^{l} L_j, \end{aligned} \tag{4.2}$$

where $L_j = \mathbb{Q}_p[x]/(p_j(x))$. Each L_j is a finite extension of $\mathbb{Q}_p$ of degree n_j since the set of left cosets

$$\{1 + (p_j(x)), x + (p_j(x)), x^2 + (p_j(x)), \dots, x^{n_j} + (p_j(x))\}$$

is a $\mathbb{Q}_p$-basis for L_j. Note that the absolute value $|\ |_{P_i}$ restricted to $\mathbb{Q}$ is the p-adic absolute value $|\ |_p$. Thus the completion of $\mathbb{Q}$ with respect to $|\ |_{P_i}$ is the completion of $\mathbb{Q}$ with respect to $|\ |_p$, and this is precisely $\mathbb{Q}_p$. Thus passing to the completion with respect to $|\ |_{P_i}$ on the left hand side of (4.2) yields the field K_{P_i}. And so, when we pass to the completion on the right hand side of (4.2), we must also obtain a field. Thus all but one of the factors on the right hand side, say, L_j must vanish upon passing to the completion with respect to $|\ |_{P_i}$. This says that for each i, $1 \le i \le g$, $K_{P_i} = L_j$ for some j, $1 \le j \le l$. It follows that the collections $\{L_i\}$ and $\{K_{P_i}\}$ coincide; $g = l$. Thus K_{P_i} is a finite field extension of $\mathbb{Q}_p$ of degree $n_i = [K_{P_i} : \mathbb{Q}_p]$. We have $n = \sum_{i=1}^{g} n_i$, as required. □

The degree $n_i = [K_{P_i} : \mathbb{Q}_p]$ is the **local degree** of K at P_i. Since $\alpha \in K$ is a root of $p(x) \in \mathbb{Q}[x]$, α is a root of $p_j(x) \in \mathbb{Q}_p[x]$ for exactly one j, $1 \le j \le l$, say j^*. Let $K_{P_{j^*}}$ be the completion corresponding to j^*. Now, $\alpha \in K_{P_{j^*}}$, and so, $K_{P_{j^*}}$ is a simple algebraic extension of $\mathbb{Q}_p$ through the evaluation homomorphism $\phi_\alpha : \mathbb{Q}_p[x] \to K_{P_{j^*}}$.

The subset

$$\hat{R}_{P_i} = \{x \in K_{P_i} : |x|_{P_i} \le 1\}$$

is a subring of K_{P_i} called the **valuation ring of** K_{P_i}. The group of units in $\hat{R}_{p_i}$ is

$$U(\hat{R}_{P_i}) = \{x \in K_{P_i} : |x|_{P_i} = 1\};$$

$\hat{R}_{P_i}$ is a local ring with maximal ideal

$$m_i = P_i\hat{R}_{P_i} = \{x \in K_{P_i} : |x|_{P_i} < 1\}.$$

Proposition 4.45. *$\hat{R}_{P_i}$ is a PID.*

Proof. It is enough to show that the maximal ideal m_i is principal. Observe that $R_{P_i} = K \cap \hat{R}_{P_i}$ and so, R_{P_i} is a subring of $\hat{R}_{P_i}$. Now, $P_iR_{P_i}$ is principal in R_{P_i} by Proposition 4.35, thus $m_i = P_i\hat{R}_{P_i}$ is a principal ideal in $\hat{R}_{P_i}$. □

An element $\pi_i \in \hat{R}_{P_i}$ with $m_i = (\pi_i)$ is called a **uniformizing parameter** for K_{P_i}. The "local form" of the factorization (4.1) is

$$(p) = m_i^{e_i} = (\pi_i)^{e_i}.$$

Note that $|\pi_i|_{P_i} = \frac{1}{p^{1/e_i}}$.

Proposition 4.46. *Every element of $x \in K_{P_i}$ can be written as*

$$x = u\pi_i^n,$$

for some unit $u \in U(\hat{R}_{P_i})$ and some $n \in Z \cup \infty$; an element $x \in \hat{R}_{P_i}$ can be written as

$$x = u\pi_i^n,$$

with $n \geq 0$.

Proof. Let $x \in K_{P_i}$. If $x = 0$, then $x = 1 \cdot \pi_i^{\infty}$. So assume that $x \neq 0$. Then $|x|_{P_i} = \frac{1}{p^{n/e_i}}$ for some integer n. Now the element $\pi_i^{-n}x$ is so that

$$|\pi_i^{-n}x|_{P_i} = |\pi_i^{-n}|_{P_i}|x|_{P_i} = p^{n/e_i}\frac{1}{p^{n/e_i}} = 1.$$

Thus $\pi^{-n}x = u$ for some unit in $\hat{R}_{P_i}$, whence $x = u\pi_i^n$. Note that $x \in \hat{R}_{P_i}$ if and only if $n \geq 0$. □

Let $x \in K_{P_i}$. The element $n \in Z \cup \infty$ for which $x = u\pi_i^n$ is the **π_i-order** of x, denoted as $\text{ord}_{\pi_i}(x)$.

Proposition 4.47. *$\hat{R}_{P_i}$ is an integral domain with Frac($\hat{R}_{P_i}$) $= K_{P_i}$.*

Proof. Clear. □

Example 4.17. Let $K = \mathbb{Q}(\mathrm{i})$; one has $\mathrm{irr}(\mathrm{i}, \mathbb{Q}) = x^2 + 1$. By Proposition 4.21, $R = Z[\mathrm{i}]$, a PID. Let p be a prime number with $p \equiv 1 \bmod 4$. Then the unique factorization of (p) is

$$(p) = P_1 P_2,$$

where $P_1 = (a - b\mathrm{i})$, $P_2 = (a + b\mathrm{i})$ for some $a, b \in Z$, cf. [Underwood (2011), §1.2]. In this case $g = 2$, $e_i = 1$ for $i = 1, 2$, and there are two prime ideals lying above (p). Thus there are exactly two extensions of $| \ |_p$ to K: $| \ |_{P_i}$ and $| \ |_{P_2}$. By Proposition 4.44, the polynomial $x^2 + 1$ factors over $\mathbb{Q}_p$ as

$$x^2 + 1 = (x - \mathrm{i})(x + \mathrm{i}).$$

Thus

$$\begin{aligned} K \otimes_{\mathbb{Q}} \mathbb{Q}_p &\cong \mathbb{Q}[x]/(x^2 + 1) \otimes_{\mathbb{Q}} \mathbb{Q}_p \\ &\cong \mathbb{Q}_p[x]/(x^2 + 1) \\ &\cong \mathbb{Q}_p[x]/(x - \mathrm{i}) \times \mathbb{Q}_p[x]/(x + \mathrm{i}) \\ &\cong \mathbb{Q}_p \times \mathbb{Q}_p. \end{aligned}$$

Note that the local degree $n_i = [K_{P_i} : \mathbb{Q}_p] = 1$ for $i = 1, 2$, and so $K_{P_i} = \mathbb{Q}_p$.

For p prime with $p \not\equiv 1 \bmod 4$, (p) remains prime in R, cf. [Underwood (2011), §1.2]. In this case $g = 1$, $e_1 = 1$. Thus there is exactly one extension of $| \ |_p$ to K, namely $| \ |_p$. The local degree $n_1 = [K_p : \mathbb{Q}_p] = 2$. Indeed, $K_p \cong \mathbb{Q}_p[x]/(x^2 + 1) \cong \mathbb{Q}_p(\mathrm{i})$; K_p is a simple algebraic extension of $\mathbb{Q}_p$ via the root $\mathrm{i} \in K_p$ and the evaluation homomorphism $\phi_{\mathrm{i}} : \mathbb{Q}_p[x] \to K_p$.

Example 4.18. Let $K = \mathbb{Q}(\zeta_{p^n})$, where ζ_{p^n} denotes a primitive p^nth root of unity. The irreducible polynomial of ζ_{p^n} over $\mathbb{Q}$ is

$$p(x) = (x^{p^{n-1}})^{p-1} + (x^{p^{n-1}})^{p-2} + \cdots + x^{p^{n-1}} + 1.$$

The ring of integers of K is $Z[\zeta_{p^n}]$. From Example 4.14, we have the unique factorization

$$(p) = (1 - \zeta_{p^n})^{p^{n-1}(p-1)}.$$

In this case, $g = 1$, $e_1 = p^{n-1}(p - 1) = [K : \mathbb{Q}]$. There is exactly one prime ideal $P = (1 - \zeta_{p^n})$ lying above (p) and so there is exactly one extension $| \ |_P$ of $| \ |_p$ to K.

The polynomial $p(x)$ remains irreducible over $\mathbb{Q}_p$; one has

$$\begin{aligned} K \otimes_{\mathbb{Q}} \mathbb{Q}_p &\cong \mathbb{Q}[x]/(p(x)) \otimes_{\mathbb{Q}} \mathbb{Q}_p \\ &\cong \mathbb{Q}_p[x]/(p(x)), \end{aligned} \tag{4.3}$$

thus $K_P \cong \mathbb{Q}_p(\zeta_{p^n})$; K_P is a simple algebraic extension of $\mathbb{Q}_p$ since $\zeta_{p^n} \in K_P$; the evaluation homomorphism is $\phi_{\zeta_{p^n}} : \mathbb{Q}_p[x] \to K_P$.

The extension of $\mathbb{Q}_p$ to K_P has local degree $[K_P : \mathbb{Q}_p] = [K : \mathbb{Q}] = p^{n-1}(p-1)$. We have $\hat{R}_P = \mathbb{Z}_p[\zeta_{p^n}]$. The unique maximal ideal $(1 - \zeta_{p^n})\mathbb{Z}_p[\zeta_{p^n}]$ is principal, generated by π; the local factorization is

$$(p) = (\pi)^{p^{n-1}(p-1)}.$$

4.8 Exercises

Exercises for §4.1

(1) Let $\alpha = 1 + \sqrt{1 + \sqrt{2}} \in \mathbb{R}$ and let $\phi_\alpha : \mathbb{Q}[x] \to \mathbb{R}$ be the evauation homomorphism.

  (a) Find the monic polynomial that generates $\ker(\phi_\alpha)$.
  (b) Compute $[\mathbb{Q}(\alpha) : \mathbb{Q}]$ and find a basis for $\mathbb{Q}(\alpha)$ over $\mathbb{Q}$.

(2) Show that $\mathbb{R}(\mathrm{i}) = \mathbb{C}$.
(3) Let $K = \mathbb{Q}(\zeta_3, \zeta_5)$.

  (a) Find an element $\eta \in \mathbb{C}$ for which $K = \mathbb{Q}(\eta)$.
  (b) Show that $4 \leq [K : \mathbb{Q}] \leq 8$.

(4) Let L be a simple algebraic extension of the field K. Prove the following formulas.

  (a) $\mathrm{Norm}_{L/K}(xy) = \mathrm{Norm}_{L/K}(x)\mathrm{Norm}_{L/K}(y)$, for $x, y \in L$.
  (b) $\mathrm{Norm}_{L/K}(r) = r^{[L:K]}$, for $r \in K$.
  (c) $\mathrm{Tr}_{L/K}(rx + y) = r\mathrm{Tr}_{L/K}(x) + \mathrm{Tr}_{L/K}(y)$, for $r \in K$, $x, y \in L$.

(5) Let $K = \mathbb{Q}(\sqrt{2} + \sqrt{3})$. Compute $\mathrm{tr}_{K/\mathbb{Q}}(\sqrt{2} + \sqrt{3})$ and $\mathrm{tr}_{K/\mathbb{Q}}(\sqrt{2})$.
(6) Let $K = \mathbb{Q}(\zeta_6)$. Compute $\mathrm{irr}(\zeta_6, \mathbb{Q})$ and calculate $\mathrm{Norm}_{K/\mathbb{Q}}(\zeta_6)$, $\mathrm{Tr}_{K/\mathbb{Q}}(\zeta_6)$.
(7) Let $K = \mathbb{Q}(\zeta_7)$ and let $\beta = \zeta_7 + \zeta_7^6 \in K$.

  (a) Compute $\mathrm{irr}(\zeta_7, \mathbb{Q}(\beta))$ and $\mathrm{irr}(\beta, \mathbb{Q})$.
  (b) Construct each embedding $K \to \mathbb{C}$ as an extension of an embedding $\mathbb{Q}(\beta) \to \mathbb{C}$.

(8) Let p be a prime number, let K be a finite extension of $\mathbb{Q}$ and suppose that $a \in K$ is not a pth power in K, that is, suppose there is no $b \in K$ for which $b^p = a$. Let $p(x) = x^p - a$.

(a) Show that the roots of $p(x)$ are

$$\sqrt[p]{a},\ \zeta_p\sqrt[p]{a},\ \zeta_p^2\sqrt[p]{a},\ldots,\zeta_p^{p-1}\sqrt[p]{a}.$$

(b) Prove that $p(x)=x^p-a$ is irreducible over K.

Exercises for §4.2

(9) Prove that $\mathbb{C}$ is the splitting field of the polynomial x^2+1 over $\mathbb{R}$. Compute $\mathrm{Gal}(\mathbb{C}/\mathbb{R})$.

(10) Let $L\subseteq\mathbb{C}$ be a simple algebraic extension of $\mathbb{R}$. Show that $\mathrm{Gal}(L/R)$ is either trivial or isomorphic to Z_2.

(11) Compute $\mathrm{Gal}(\mathbb{Q}(\zeta_5):\mathbb{Q})$, $\mathrm{Gal}(\mathbb{Q}(\zeta_5):\mathbb{Q}(\zeta_5+\zeta_5^4))$ and $\mathrm{Gal}(\mathbb{Q}(\zeta_5+\zeta_5^4):\mathbb{Q})$.

(12) Let $\alpha=1+\sqrt{1+\sqrt{2}}\in\mathbb{R}$. Let K be the splitting field of $\mathrm{irr}(\alpha,\mathbb{Q})$. Compute $\mathrm{Gal}(K:\mathbb{Q})$.

(13) Let K be the splitting field of the polynomial $x^4-3\in\mathbb{Q}[x]$.

(a) Compute $\mathrm{Gal}(K/\mathbb{Q})$.
(b) Find all of the subfields of K.

(14) Let K be a field $\mathbb{Q}\subseteq K\subseteq\mathbb{C}$. Let L be the splitting field of the monic irreducible polynomial $p(x)\in K[x]$ of degree n. Prove that $|\mathrm{Gal}(L/K)|\le n!$.

Exercises for §4.3

(15) Let $K=\mathbb{Q}(\alpha)$ and let γ,β be elements of K that are integral over Z. Prove that $\gamma+\beta$ is integral over Z.

(16) Prove that the integral closure of Z in $\mathbb{Q}$ is Z.

(17) Let $m=-5$ and let $\mathbb{Q}(\sqrt{-5})$ be the imaginary quadratic field extension of $\mathbb{Q}$.

(a) Compute the ring of integers R of $\mathbb{Q}(\sqrt{-5})$.
(b) Show that R is not a PID.

(18) Let p be a prime number and let $n\ge 1$ be an integer. In this exercise, we will prove that the ring of integers of $K=\mathbb{Q}(\zeta_{p^n})$ is $R=Z[\zeta_{p^n}]$. This is not that hard if we have the following fact: the irreducible polynomial of ζ_{p^n} over $\mathbb{Q}$ is

$$p(x)=(x^{p^{n-1}})^{p-1}+(x^{p^{n-1}})^{p-2}+\cdots+(x^{p^{n-1}})^2+x^{p^{n-1}}+1.$$

The distinct zeros of $p(x)$ are $\{\zeta_{p^n}^i\}$, $1\le i\le p^n-1$, $\gcd(i,p)=1$. The embeddings $K\to\mathbb{C}$ are of the form $\tau_i:\zeta_{p^n}\mapsto\zeta_{p^n}^i$, $1\le i\le p^n-1$,

$\gcd(i,p) = 1$. Now with this information, prove that $R = Z[\zeta_{p^n}]$ by completing the following steps.

(a) First, show that $R(1-\zeta_{p^n}) \cap Z = pZ$.
(b) Use Part (a) to show that $\mathrm{Tr}_{K/\mathbb{Q}}(R(1-\zeta_{p^n})) \subseteq pZ$.
(c) Next, use the method of Proposition 4.22 to establish $R = Z[\zeta_{p^n}]$.

(19) Compute the number of fundamental units in the ring of integers of $\mathbb{Q}(\zeta_{p^n})$.
(20) Find a fundamental unit for the ring of integers of $\mathbb{Q}(\sqrt{3})$.
(21) Find a fundamental unit for the ring of integers of $\mathbb{Q}(\sqrt[3]{2})$.
(22) Find a fundamental unit for the ring of integers of $\mathbb{Q}(\zeta_p)$.
(23) Find a non-trivial unit in the group ring ZC_7.

Exercises for §4.4

(24) Let m be a square-free positive integer with $m \equiv 1 \bmod 4$. Let $K = \mathbb{Q}(\sqrt{m})$ with ring of integers R. Let $B : K \times K \to \mathbb{Q}$ be the bilinear form defined as $B(x,y) = \mathrm{Tr}_{K/\mathbb{Q}}(xy)$. Compute $\mathrm{disc}(R)$ with respect to the bilinear form B.
(25) Let $K = \mathbb{Q}(\zeta_{p^2})$ with ring of integers $R = Z[\zeta_{p^2}]$. Compute $\mathrm{disc}(R)$ with respect to the bilinear form $B : K \times K \to \mathbb{Q}$, $B(x,y) = \mathrm{Tr}_{K/\mathbb{Q}}(xy)$.
(26) Let K be a simple algebraic extension of $\mathbb{Q}$ with ring of integers R. Let A, B be non-zero ideals of R with $A = AB$. Prove that $B = R$.
(27) Let K be a simple algebraic extension of $\mathbb{Q}$ with ring of integers R. Let J be a non-zero ideal of R. Prove that $J \cap Z^+$ is non-empty.

Exercises for §4.5

(28) Prove that a PID is a Dedekind domain.
(29) Let K be a simple algebraic extension of $\mathbb{Q}$ with ring of integers R. Let I, J be fractional ideals over R. Prove the following statements.

(a) $I + J$ is a fractional ideal,
(b) IJ is a fractional ideal,
(c) $I \cap J$ is a fractional ideal.

(30) Prove that a fractional ideal J over R with $J \subseteq R$ is an ideal of R.
(31) Let I be a fractional ideal over R. Show there exists an element $a \in R$ for which $aI \subseteq R$.

(32) Suppose J is a fractional ideal with $RJ \subset R$. Show that J can be written in the form $J = IP_1^{e_1}P_2^{e_2}\cdots P_k^{e_k}$, where I is a fractional ideal and P_i are prime ideals. Is the collection $\{P_i\}$ unique?

(33) Prove Proposition 4.39.

(34) Let K be a simple algebraic extension of $\mathbb{Q}$ with ring of integers R. Prove the following: R is a PID if and only if the class group of R is trivial.

(35) Find an example of a simple algebraic extension of $\mathbb{Q}$ for which the class group is non-trivial.

Exercises for §4.6

(36) Let R be the ring of integers of $K = \mathbb{Q}(\alpha)$. Let J be a non-trivial proper ideal of R. Show that the factorization

$$J = P_1^{e_1}P_2^{e_2}\cdots P_k^{e_k},$$

is unique up to a re-ordering of the factors $P_i^{e_i}$.

(37) Give the unique factorization of the ideal (100) in Z.

(38) Obtain the unique factorization of the ideal (2) in $Z[\zeta_3]$. Is $Z[\zeta_3]/(2)$ a field? Is $Z[\zeta_3]/(2)$ an integral domain?

Exercises for §4.7

(39) Let $K = \mathbb{Q}(\sqrt{2})$.

(a) Compute the number of extensions of $|\ |_2$ to K.
(b) Compute the number of extensions of $|\ |_3$ to K.

(40) Let K be the splitting field of the polynomial $x^3 - 3$ over $\mathbb{Q}$. Compute the number of extensions of $|\ |_2$ to K.

(41) Let $K = \mathbb{Q}(\sqrt[3]{3})$. Compute the number of extensions of $|\ |_2$ to K.

(42) Let K be a simple algebraic extension of $\mathbb{Q}$ with ring of integers R. Let P be a prime ideal of R lying above (p), that is, suppose that P occurs in the unique factorization of (p). Prove that $\hat{R}_P \cap K = R_P$.

(43) Let $K = \mathbb{Q}(\mathrm{i})$ with $R = Z[\mathrm{i}]$.

(a) Decompose the ideal (13) of R into a product of prime ideals of R.
(b) For each prime ideal P_i in the factorization of Part (a) compute $|13|_{P_i}$.

(44) Let $K = \mathbb{Q}(\mathrm{i})$ with $R = Z[\mathrm{i}]$.

(a) Decompose (2) into a product of prime ideals of R.
(b) Verify the decomposition $(30) = (1+\mathrm{i})^2(3)(2-\mathrm{i})(2+\mathrm{i})$. Compute $|30|_{(1+\mathrm{i})}$ and $|30|_{(2+\mathrm{i})}$.

(45) Let p be a prime number, let $n \geq 1$ and let $K = \mathbb{Q}(\zeta_{p^n})$ where ζ_{p^n} is a primitive p^nth root of unity. Let R be the ring of integers of K, $P = (1 - \zeta_{p^n})$. Compute $|1 - \zeta_p|_P$ and $|1 + \zeta_p|_P$.

Questions for Further Study

(1) Referring to Example 4.16, let $K = \mathbb{Q}(\sqrt{-5})$ with ring of integers $R = Z[\sqrt{-5}]$.

(a) Prove that the class number of R is 2.
(b) Let $L = K(\mathrm{i})$. Show that the ring of integers of L is $S = R[(\mathrm{i} + \sqrt{-5})/2]$.
(c) Prove that the L is the Hilbert Class Field of K.

(2) Let $K = \mathbb{Q}(\sqrt{3})$ and let α be a zero of $p(x) = x^2 + \sqrt{3}x + 1 \in K[x]$.

(a) Show that the ring of integers of $L = K(\alpha)$ is $S = Z[\sqrt{3}, \alpha]$.
(b) Show that every non-zero prime ideal Q of R is unramified in S.
(c) Is L the Hilbert Class Field of K?

Chapter 5

Finite Fields

In this final chapter we introduce finite fields. Using the concept of invented roots we give an analog of the Fundamental Theorem of Algebra: we show that if F is any field and $f(x) \in F[x]$, then there exists a field extension E/F so that $f(x)$ factors into linear factors in $E[x]$. Necessarily, E contains all of the zeros of $f(x)$. In the case that $F = Z_p$, p a prime number, and $f(x) = x^{p^n} - x$, $n \geq 1$, there is a field extension E/Z_p that contains p^n zeros of $f(x)$. These zeros are distinct and form a subfield of E called the Galois field of p^n elements, $\mathrm{GF}(p^n)$. We have $Z_p \subseteq \mathrm{GF}(p^n) \subseteq E$.

We consider polynomials $f(x)$ over $\mathrm{GF}(p^n)$, define the order of $f(x)$, denoted as $\mathrm{order}(f(x))$, and show that if $f(x)$ is irreducible of degree k, then $\mathrm{order}(f(x))$ is the order of any zero α of $f(x)$ in the group of units of the simple algebraic extension $\mathrm{GF}(p^n)(\alpha) = \mathrm{GF}(p^{nk})$. We prove that if $f(x)$ is a primitive polynomial over $\mathrm{GF}(p^n)$ of degree k, then $\mathrm{order}(f(x))$ has maximal value $p^{nk}-1$, that is, if α is any zero of $f(x)$, then $\langle\alpha\rangle = \mathrm{GF}(p^{nk})^\times$.

We next consider linearly recursive sequences over an arbitrary field K and give some examples, including the Fibonacci sequence. We specialize to sequences over the Galois field $\mathrm{GF}(p^m)$. We prove that if $\{s_n\}$ is a kth-order linearly recursive sequence over $\mathrm{GF}(p^m)$ with primitive characteristic polynomial $f(x)$, then $\{s_n\}$ has maximal period $p^{mk} - 1$.

5.1 Invented Roots

In §4.1 we stated the following: if K is a subfield $\mathbb{C}$, and $f(x) \in K[x]$, then there exist a zero of $f(x)$ in $\mathbb{C}$. This is the Fundamental Theorem of Algebra. But what can we say if F is an arbitrary field not necessarily contained in $\mathbb{C}$? In this section, we show that if F is any field and $f(x)$ is an irreducible polynomial over F, then there exists a field extension E/F that contains a zero of $f(x)$; this zero is "invented" as the left coset $x + (f(x))$ in $F[x]/(f(x))$. Analogous to the field extension $\mathbb{C}/K$, we construct a field extension E/F so that $f(x)$ factors into linear factors over E.

* * *

Let F be any field and let $q(x)$ be an irreducible polynomial in $F[x]$.

Proposition 5.1. *There exists a field extension E/F that contains a zero of $q(x)$.*

Proof. Since $q(x)$ is irreducible, $(q(x))$ is a maximal ideal of $F[x]$. Thus $E = F[x]/(q(x))$ is a field. We identify F with the collection of cosets $\{r + (q(x)) : r \in F\}$, and in this way, $F \subseteq E$. Now let $\alpha = x + (q(x)) \in E$. One has $q(\alpha) = q(x + (q(x))) = (q(x))$ and so $q(\alpha) = 0$ in E. (Note: the evaluation of q at the coset $x + (q(x))$ makes sense since we are viewing the quotient ring $F(x)/(q(x))$ as an F-module in the standard way.) $\square$

We call the coset α an **invented root** of $q(x)$.

Proposition 5.2. *Let $f(x) \in F[x]$. Then there exists a field extension E/F so that $f(x)$ factors into a product of linear factors in $E[x]$.*

Proof. We prove this by induction on $d = \deg(f(x))$.
The trivial case: $d = 1$. In this case $f(x) = ax + b \in F[x]$, so we may take $E = F$.
The induction hypothesis. We assume that the proposition is true for $f(x)$ with $\deg(f(x)) = d - 1$. By Proposition 2.30, $F[x]$ is a UFD, and so, $f(x)$ factors into a product of irreducible polynomials

$$f(x) = q_1(x)q_2(x)q_3(x)\cdots q_k(x).$$

If $\deg(q_i(x)) = 1$ for $i = 1, 2, \ldots, k$, then we can take $E = F$. Else, let j be the smallest index with $\deg(q_j(x)) > 1$. By Proposition 5.1 there exists an invented root α of $q_j(x)$ in some field extension L/F. Over L, $f(x)$ factors as

$$f(x) = q_1(x)q_2(x)\cdots q_{j-1}(x)(x - \alpha)r(x)q_{j+1}\cdots q_k(x),$$

for $r(x) \in L[x]$.

Put

$$g(x) = q_1(x)q_2(x)\cdots q_{j-1}(x)r(x)q_{j+1}\cdots q_k(x).$$

Then $\deg(g(x)) = d - 1$. By the induction hypothesis there exists a field extension E/L so that that $g(x)$ factors into a product of linear factors in $E[x]$. Since $f(x) = g(x)(x - \alpha)$, $f(x)$ factors into a product of linear factors in $E[x]$. □

Proposition 5.2 says that E/F contains d roots of $f(x)$ which may or may not be distinct. These are the only possible zeros of $f(x)$ since a degree d polynomial over a field can have at most d roots in the field (Proposition 2.6).

5.2 Finite Fields

Let p be a prime number and let $n \geq 1$ be an integer. In this section we prove the existence of a field with exactly p^n elements. We show that there is essentially only one field with p^n elements, which we call the Galois field $\mathrm{GF}(p^n)$ of order p^n. If $f(x)$ is an irreducible polynomial of degree k over $\mathrm{GF}(p^n)$ and α is an invented root of $f(x)$, then $\mathrm{GF}(p^n)(\alpha)$ is the Galois field $\mathrm{GF}(p^{nk})$. We define the order of a polynomial $f(x) \in \mathrm{GF}(p^n)[x]$; a primitive polynomial over $\mathrm{GF}(p^n)$ is an irreducible polynomial whose order is maximal. We show that if $f(x)$ is primitive over $\mathrm{GF}(p^n)$ and α is a root of $f(x)$, then α generates the cyclic group $\mathrm{GF}(p^{nk})^\times$.

* * *

A **finite field** is a field with a finite number of elements. By §2.8, Exercise 3, we know that Z_p is a finite field with exactly p elements for each prime number p. It turns out that the number of elements in any finite field is always a power of a prime number.

Proposition 5.3. *Let F be a field with a finite number of elements. Then F is isomorphic to a simple algebraic extension of Z_p for some prime number p. Consequently, $|F| = p^n$ where n is the degree of the simple algebraic extension of Z_p.*

Proof. Since F is finite, $r = \mathrm{char}(F) > 0$, hence by Corollary 2.6, F contains a subring B isomorphic to Z_r. Henceforth, we identify B with Z_r. Since F is a field, r must be a prime number p, hence F contains the field Z_p. As F is finite, it is certainly a finite dimensional vector space over Z_p, with scalar multiplication $Z_p \times F \to F$ given by multiplication in F. Thus $F = \underbrace{Z_p \oplus Z_p \oplus \cdots \oplus Z_p}_{n}$, where $n = \dim(F)$, whence $|F| = p^n$.

By Proposition 2.15, the group of units of F, $F^\times$, is cyclic of order p^n-1; let α generate $F^\times$. Thus $\alpha^{p^n-1}=1$, that is, α is a root of the polynomial $r(x)=x^{p^n-1}-1\in Z_p[x]$. We have the evaluation homomorphism $\phi_\alpha : Z_p[x]\to F$ with $\ker(\phi_\alpha)=(q(x))$, where $q(x)$ is the polynomial of smallest degree m in $Z_p[x]$ for which α is a root. Then $q(x)=\mathrm{irr}(\alpha,Z_p)$, so that $Z_p[x]/(q(x))\cong Z_p(\alpha)$ is a simple algebraic extension of Z_p.

Let $\psi : Z_p[x]/(q(x))\to F$ be defined by $f(x)+(q(x))\mapsto f(\alpha)$. Since $g(x)-f(x)\in(q(x))$ implies that $g(\alpha)=f(\alpha)$, ψ is well-defined on cosets. Clearly, ψ is onto, and since $Z_p[x]/(q(x))$ is a field, ψ is 1-1. Hence ψ is an isomorphism of fields. The set $\{1,\overline{x},\overline{x}^2,\dots,\overline{x}^{m-1}\}$ is a Z_p-basis for $Z_p[x]/((q(x))$ and so, $|F|=p^m$. But we already know that $|F|=p^n$, hence $m=n$. □

Let p be a prime number and let $n\geq 1$ be an integer. In the case $n=1$, there exists a field with exactly $p^1=p$ elements, namely, Z_p. We now show that this is true for $n>1$: we show that there exists a field with exactly p^n elements for all $n\geq 1$. Consider $f(x)=x^{p^n}-x\in Z_p[x]$. Then by Proposition 5.2 there exists a field extension E/Z_p which contains p^n zeros of $f(x)$ (counting multiplicities). These constitute all of the zeros of $f(x)$.

Proposition 5.4. *The zeros of $f(x)=x^{p^n}-x$ in E are distinct.*

Proof. Let $F=\{\alpha_i\}$, $1\leq i\leq p^n$ be the set of roots of $f(x)$, and suppose that some root α_i has multiplicity ≥ 2. Then $f'(\alpha_i)=0$. But this is impossible since $f'(x)=-1$ in $Z_p[x]$. □

Proposition 5.5. *Let $F=\{\alpha_i\}$, $1\leq i\leq p^n$, be the set of roots of $f(x)=x^{p^n}-x$. Then F is a field, with operations induced from E.*

Proof. By Corollary 2.1 the elements of Z_p are roots of $f(x)$. Thus $Z_p\subseteq F$ and $\mathrm{char}(F)=p$. Let $\alpha_i,\alpha_j\in F$. Then $(\alpha_i+\alpha_j)^{p^n}=\alpha_i^{p^n}+\alpha_j^{p^n}=\alpha_i+\alpha_j$. (The first equality follows from §2.8, Exercise 77.) Thus F is closed under addition. Moreover, $(-\alpha_i)^{p^n}=(-1)^{p^n}\alpha_i^{p^n}=-\alpha_i$, since $(-1)^{p^n}\equiv -1 \bmod p$ by Corollary 2.1. Hence F is an additive subgroup of E. Also, $(\alpha_i\alpha_j)^{p^n}=\alpha_i^{p^n}\alpha_j^{p^n}=\alpha_i\alpha_j$, so that F is closed under multiplication. Thus F is a commutative ring with unity. For $\alpha_i\in F$ non-zero, $\alpha_i^{-1}\in E$. But also, $(\alpha_i^{-1})^{p^n}=\alpha_i^{-1}$. Thus F is a field. □

By Proposition 5.5, there is a field F of order p^n, consisting of the p^n distinct roots of $x^{p^n}-x$. The following proposition shows that there is essentially only one finite field of order p^n.

Proposition 5.6. *Let F_1, F_2 be finite fields of order p^n. Then $F_1 \cong F_2$.*

Proof. We show that $F_1 \cong F$, where F is the field constructed in Proposition 5.5 consisting of the roots of $x^{p^n} - x \in Z_p[x]$. By Proposition 5.3, F_1 is isomorphic to a simple algebraic extension of Z_p, $Z_p[x]/(q(x)) \cong Z_p(\alpha)$, where $\deg(q(x)) = n$, and α is a root of both $q(x)$ and $x(x^{p^n-1} - 1) = x^{p^n} - x$. Hence $\alpha \in F$. Define $\phi : Z_p[x]/(q(x)) \to F$ by the rule $f(x)+(q(x)) \mapsto f(\alpha)$. Then is well-defined on cosets (ϕ is a function). Moreover, ϕ is 1-1 since $Z_p[x]/(q(x))$ is a field, and onto since both $Z_p[x]/(q(x))$ and F have the same number of elements. Thus ϕ is an isomorphism. It follows that $F_1 \cong F$. Similarily, $F_2 \cong F$, which proves the result. □

The unique (up to isomorphism) finite field of order p^n is called the **Galois field** of order p^n and is denoted by $\mathrm{GF}(p^n)$.

We next discuss polynomials over the Galois field $\mathrm{GF}(p^n)$.

Proposition 5.7. *Let $f(x) \in GF(p^n)[x]$ with $\deg(f(x)) = k \geq 1$. Assume that $f(0) \neq 0$. Then there exists an integer t, $1 \leq t \leq p^{nk} - 1$, for which $f(x) \mid x^t - 1$.*

Proof. First note that the quotient ring $\mathrm{GF}(p^n)[x]/(f(x))$ contains $p^{nk} - 1$ left cosets other than $(f(x))$. The collection $\{x^i + (f(x)) : i = 0, 1, 2, \ldots, p^{nk} - 1\}$ is a set of p^{nk} left cosets not containing $(f(x))$. Thus

$$x^i + (f(x)) = x^j + (f(x)),$$

for some i, j, $0 \leq i < j \leq p^{nk} - 1$. Since $f(0) \neq 0$, $\gcd(x, f(x)) = 1$. Thus by Proposition 2.21 there exists polynomials $r(x), s(x) \in \mathrm{GF}(p^n)[x]$ so that

$$xr(x) + f(x)s(x) = 1.$$

Consequently, $x^i + (f(x))$ is a unit in $\mathrm{GF}(p^n)[x]/(f(x))$. Set $t = j - i$. It follows that $x^t \in 1 + (f(x))$ and hence $f(x) \mid x^t - 1$ with $1 \leq t \leq p^{nk} - 1$. □

The smallest positive integer e for which $f(x) \mid x^e - 1$ is the **order of** $f(x)$, denoted as $\mathrm{order}(f(x))$. For example, over $\mathrm{GF}(2)$, $x^4 + x + 1$ has order 15 and $x^4 + x^3 + x^2 + x + 1$ has order 5; over $\mathrm{GF}(9) = \mathrm{GF}(3)(\alpha)$, $\alpha^2 + 1 = 0$, $x - \alpha$ has order 4 since

$$x^4 - 1 = (x - 1)(x + 1)(x - \alpha)(x + \alpha)$$

over $\mathrm{GF}(9)$.

Proposition 5.8. *Let $f(x)$ be an irreducible polynomial in $GF(p^n)[x]$ of degree k. Then $f(x)$ divides $x^{p^{nm}} - x$ if and only if $k \mid m$.*

Proof. Suppose that $f(x)$ divides $x^{p^{nm}} - x$. By Proposition 5.1, there exists a field extension $E/\mathrm{GF}(p^n)$ that contains a zero α of $f(x)$. Thus $\alpha^{p^{nm}} - \alpha = 0$, and so, $\alpha \in \mathrm{GF}(p^{nm})$. Moreover, the elements of $\mathrm{GF}(p^n)$ are precisely the zeros of $x^{p^n} - x$ and any zero of $x^{p^n} - x$ is also a zero of $x^{p^{nm}} - x$. Thus, $F = \mathrm{GF}(p^n)(\alpha)$ is a subfield of $\mathrm{GF}(p^{nm})$. Note that F has p^{nk} elements.

Now let β be a generator of the cyclic group $\mathrm{GF}(p^{nm})^\times$ and let $q(x)$ be the irreducible polynomial of β over F. Let $t = \deg(q(x))$. Then $F(\beta) = \mathrm{GF}(p^{nm})$. Now $F(\beta)$ has p^{nkt} elements and so $p^{nkt} = p^{nm}$. Hence $kt = m$, that is, $k \mid m$.

For the converse, suppose that $k \mid m$. Let α be a zero of $f(x)$ in some extension field $E/\mathrm{GF}(p^n)$. Then $\mathrm{GF}(p^n)(\alpha)$ is a field with p^{nk} elements, and α satisfies the relation $\alpha^{p^{nk}} - \alpha = 0$. Let s be so that $ks = m$. Then $\alpha^{p^{nks}} - \alpha = 0$, hence, $\alpha^{p^{nm}} - \alpha = 0$. Consequently, α is a root of $x^{p^{nm}} - x$. It follows that $x^{p^{nm}} - x$ is in $(f(x))$, the kernel of the evaluation homomorphism $\phi_\alpha : \mathrm{GF}(p^n)[x] \to E$, and so $f(x) \mid x^{p^{nm}} - x$. □

Proposition 5.9. *Let $f(x)$ be an irreducible polynomial in $GF(p^n)[x]$ of degree k. Let α be a zero of $f(x)$ in an extension field $E/GF(p^n)$. Then the zeros of $f(x)$ are of the form $\alpha, \alpha^{p^n}, \alpha^{p^{2n}}, \ldots, \alpha^{p^{(k-1)n}}$. Moreover, the zeros are distinct.*

Proof. We already know that $\alpha \in E$ is one zero of $f(x)$. Since the characteristic of $\mathrm{GF}(p^n)$ is p,

$$f(\alpha^{p^{jn}}) = f(\alpha)^{p^{jn}} = 0,$$

for $1 \le j \le k-1$. Now suppose that $\alpha^{p^{an}} = \alpha^{p^{bn}}$ for integers $0 \le a < b \le k-1$. Then

$$\alpha^{p^{n(k+a-b)}} = (\alpha^{p^{na}})^{p^{n(k-b)}} = (\alpha^{p^{nb}})^{p^{n(k-b)}} = \alpha^{p^{nk}} = \alpha.$$

Thus, $f(x)$ divides $x^{p^{n(k+a-b)}} - x$, and so by Proposition 5.8, k divides $k + a - b$, which is impossible. It follows that the zeros of $f(x)$ are distinct. □

Proposition 5.10. *Let $f(x)$ be an irreducible polynomial in $GF(p^n)[x]$ of degree k. Let α be a zero of $f(x)$ in an extension field $E/GF(p^n)$. The smallest field extension containing all of the zeros of $f(x)$ is $GF(p^n)(\alpha)$, which is isomorphic to the Galois field $GF(p^{nk})$.*

Proof. By Proposition 5.9, $\mathrm{GF}(p^n)(\alpha)$ contains all of the zeros of $f(x)$. Clearly, $|\mathrm{GF}(p^n)(\alpha)| = p^{nk}$. Now, by Proposition 5.6, $\mathrm{GF}(p^n)(\alpha) \cong \mathrm{GF}(p^{nk})$. □

Let $f(x)$ be an irreducible polynomial in $\mathrm{GF}(p^n)[x]$ of degree $k \geq 1$, and let α be a zero of $f(x)$. As we have seen in Proposition 5.10, $\mathrm{GF}(p^n)(\alpha)$ contains all of the roots of $f(x)$. The following proposition computes $\mathrm{order}(f(x))$.

Proposition 5.11. *With $f(x)$ as above, order($f(x)$) equals the order of any root of $f(x)$ in the group of units of GF(p^n)(α).*

Proof. Observe that $\mathrm{GF}(p^n)(\alpha)^\times$ is cyclic of order $p^{nk} - 1$, generated by some element β. Put $\alpha = \beta^l$ for some integer l. Now a typical zero of $f(x)$ can be written $\beta^{lp^{mn}}$ for $0 \leq m \leq k-1$. By Proposition 1.34,

$$|\langle \beta^{lp^{mn}} \rangle| = \frac{p^{nk} - 1}{\gcd(p^{nk} - 1, lp^{mn})}.$$

Since $\gcd(p^{nk} - 1, p^{mn}) = 1$, the right hand side above only depends on l, and so each zero of $f(x)$ has the same order.

Let e be the order of α in $\mathrm{GF}(p^n)(\alpha)^\times$ (the smallest positive integer e so that $\alpha^e = 1$). Then α is a zero of $x^e - 1$. Thus $x^e - 1 \in (f(x))$ since $(f(x))$ is the kernel of the evaluation homomorphism $\phi_\alpha : \mathrm{GF}(p^n)[x] \to E$. Thus $f(x) \mid x^e - 1$. It follows that $e = \mathrm{order}(f(x))$. □

An irreducible polynomial $f(x)$ of degree k in $\mathrm{GF}(p^n)[x]$ is **primitive** if $\mathrm{order}(f(x)) = p^{nk} - 1$. Equivalently, an irreducible, degree k polynomial $f(x)$ is primitive if there exists a root α of $f(x)$ for which $\langle \alpha \rangle = \mathrm{GF}(p^n)(\alpha)^\times$. For example, in GF(2), every irreducible polynomial of degree $k = 3, 5, 7$ is primitive since the group of units $\mathrm{GF}(2^k)^\times$ has prime order for $k = 3, 5, 7$.

We give some examples of Galois fields of order p^n and provide examples of primitive and non-primitive polynomials.

Example 5.1. Let $p = 5$, $n = 1$. The Galois field GF(5) is the finite field Z_5; Z_5 contains all of the distinct roots of $x^5 - x \in Z_5[x]$ by Corollary 2.1. Indeed, over Z_5,

$$x^5 - x = x(x-1)(x-2)(x-3)(x-4).$$

Observe that 2 has order 4 in $Z_5^\times$ and 4 has order 2 in $Z_5^\times$. Hence $\mathrm{order}(x - 2) = 4$ and $\mathrm{order}(x-4) = 2$; $x - 2$ is a primitive polynomial over Z_5.

Example 5.2. Let $p = 3$, $n = 2$. Then GF(9) consists of all of the roots of $x^9 - x$. In the UFD $Z_3[x]$, $x^9 - x$ factors into irreducible elements as:

$$x^9 - x = x(x-1)(x+1)(x^2+1)(x^2-x-1)(x^2+x-1).$$

We take α to be a root of $x^2 + 1$ (the other root is $\alpha^3 = 2\alpha$). Thus $\mathrm{GF}(9) \cong Z_3[x]/(x^2+1) \cong Z_3(\alpha)$. A Z_3-basis for $Z_3(\alpha)$ is $\{1, \alpha\}$. The 9 elements of GF(9) are thus:

$$\begin{aligned} 0 &= 0 \cdot 1 + 0 \cdot \alpha, \\ \alpha &= 0 \cdot 1 + 1 \cdot \alpha, \\ 2\alpha &= 0 \cdot 1 + 2 \cdot \alpha, \\ 1 &= 1 \cdot 1 + 0 \cdot \alpha, \\ 1+\alpha &= 1 \cdot 1 + 1 \cdot \alpha, \\ 1+2\alpha &= 1 \cdot 1 + 2 \cdot \alpha, \\ 2 &= 2 \cdot 1 + 0 \cdot \alpha, \\ 2+\alpha &= 2 \cdot 1 + 1 \cdot \alpha, \\ 2+2\alpha &= 2 \cdot 1 + 2 \cdot \alpha. \end{aligned}$$

Both α and α^3 have order 4 in $\mathrm{GF}(9)^\times$, and so $\mathrm{order}(x^2+1) = 4$. Thus $x^2 + 1$ is a non-primitive polynomial over GF(3). On the other hand, the root β of $x^2 - x - 1 \in \mathrm{GF}(3)[x]$ has order 8 in $\mathrm{GF}(9)^\times$, and so $x^2 - x - 1$ is a primitive polynomial over GF(3).

Example 5.3. In this example, we take $\mathrm{GF}(9) = \mathrm{GF}(3)(\alpha)$, $\alpha^2 + 1 = 0$, as our base field. Let $f(x) = x^2 + x + \beta \in \mathrm{GF}(9)[x]$ with β as in Example 5.2. Then one checks directly that $f(x)$ is irreducible over GF(9). By Proposition 5.9, the (distinct) roots of $f(x)$ are γ, γ^9; $\mathrm{GF}(9)(\gamma) = \mathrm{GF}(81)$. We have

$$(x - \gamma)(x - \gamma^9) = x^2 + x + \beta,$$

so that $\gamma^{10} = \beta$. Since β has order 8 in GF(9), γ has order 80 in $\mathrm{GF}(81)^\times$. Thus $f(x)$ is primitive over GF(9).

The finite fields $\mathrm{GF}(2^n)$, $n \geq 1$, can be applied to computer science since their base field $\mathrm{GF}(2) = \{0, 1\}$ represents the collection of binary digits (bits).

Example 5.4. Consider GF(16). The polynomial $x^{16} - x \in Z_2[x]$ factors into irreducibles as

$$x(x+1)(x^2+x+1)(x^4+x+1)(x^4+x^3+1)(x^4+x^3+x^2+x+1) \tag{5.1}$$

So GF(16) can be constructed as $\mathrm{GF}(2)[x]/(x^4+x+1) = \mathrm{GF}(2)(\alpha)$, where α is a root of x^4+x+1. A GF(2)-basis for GF(16) is $\{1, \alpha, \alpha^2, \alpha^3\}$ and so the 16 elements are

$$\begin{aligned}
0 &= 0\cdot 1 + 0\cdot\alpha + 0\cdot\alpha^2 + 0\cdot\alpha^3,\\
\alpha^3 &= 0\cdot 1 + 0\cdot\alpha + 0\cdot\alpha^2 + 1\cdot\alpha^3,\\
\alpha^2 &= 0\cdot 1 + 0\cdot\alpha + 1\cdot\alpha^2 + 0\cdot\alpha^3,\\
\alpha^2+\alpha^3 &= 0\cdot 1 + 0\cdot\alpha + 1\cdot\alpha^2 + 1\cdot\alpha^3,\\
\alpha &= 0\cdot 1 + 1\cdot\alpha + 0\cdot\alpha^2 + 0\cdot\alpha^3,\\
\alpha+\alpha^3 &= 0\cdot 1 + 1\cdot\alpha + 0\cdot\alpha^2 + 1\cdot\alpha^3,\\
\alpha+\alpha^2 &= 0\cdot 1 + 1\cdot\alpha + 1\cdot\alpha^2 + 0\cdot\alpha^3,\\
\alpha+\alpha^2+\alpha^3 &= 0\cdot 1 + 1\cdot\alpha + 1\cdot\alpha^2 + 1\cdot\alpha^3,\\
1 &= 1\cdot 1 + 0\cdot\alpha + 0\cdot\alpha^2 + 0\cdot\alpha^3,\\
1+\alpha^3 &= 1\cdot 1 + 0\cdot\alpha + 0\cdot\alpha^2 + 1\cdot\alpha^3,\\
1+\alpha^2 &= 1\cdot 1 + 0\cdot\alpha + 1\cdot\alpha^2 + 0\cdot\alpha^3,\\
1+\alpha^2+\alpha^3 &= 1\cdot 1 + 0\cdot\alpha + 1\cdot\alpha^2 + 1\cdot\alpha^3,\\
1+\alpha &= 1\cdot 1 + 1\cdot\alpha + 0\cdot\alpha^2 + 0\cdot\alpha^3,\\
1+\alpha+\alpha^3 &= 1\cdot 1 + 1\cdot\alpha + 0\cdot\alpha^2 + 1\cdot\alpha^3,\\
1+\alpha+\alpha^2 &= 1\cdot 1 + 1\cdot\alpha + 1\cdot\alpha^2 + 0\cdot\alpha^3,\\
1+\alpha+\alpha^2+\alpha^3 &= 1\cdot 1 + 1\cdot\alpha + 1\cdot\alpha^2 + 1\cdot\alpha^3.
\end{aligned}$$

In fact, $f(x) = x^4+x+1$ is a primitive polynomial over GF(2): From the factorization (5.1), $f(x)$ is irreducible. By Proposition 5.11, $\mathrm{order}(f(x)) = 3$ or 5 or 15. But clearly, $f(x) \nmid x^3-1$ and $f(x) \nmid x^5-1$, thus $\mathrm{order}(f(x)) = 15$ which says that $f(x)$ is primitive.

GF(16) is the field consisting of all possible half-bytes (strings of 0's and 1's of length 4). The addition is given by bit-wise addition modulo 2 and the multiplication is induced by the relation $\alpha^4 = 1+\alpha$. For example,

$$0110 + 1100 = 1010,$$

and

$$0110 \cdot 1001 = 1100,$$

since $(\alpha+\alpha^2)(1+\alpha^3) = 1+\alpha$.

The analogous bit-string multiplication in $\mathrm{GF}(2^8) = \mathrm{GF}(256)$ is used in the construction of the symmetric key cryptosystem AES (the Advanced Encryption Standard), see [Mao (2004), §7.7].

5.3 Linearly Recursive Sequences

In this final section we define kth-order linearly recursive sequences over an arbitrary field K and give some examples, including the arithmetic sequence, the geometric sequence and the Fibonacci sequence. We specialize to homogeneous sequences and construct the matrix A of a homogeneous sequence and the characteristic polynomial of the sequence. From the Cayley-Hamilton Theorem, we deduce that the minimal polynomial of A is the characteristic polynomial of A.

We next consider kth-order linearly recursive sequences $\{s_n\}$ over the Galois field $\mathrm{GF}(p^m)$. We prove that if the characteristic polynomial $f(x)$ of $\{s_n\}$ is primitive over $\mathrm{GF}(p^m)$, then $\{s_n\}$ has maximal period $p^{mk} - 1$.

* * *

Definition 5.1. Let K be a field and let $k > 0$ be a positive integer. A **kth-order linearly recursive sequence in K** is a sequence $\{s_n\}$ for which

$$s_{n+k} = a_{k-1}s_{n+k-1} + a_{k-2}s_{n+k-2} + \cdots + a_0 s_n + a \tag{5.2}$$

for some elements $a, a_0, a_1, a_2, \ldots, a_{k-1} \in K$ and all $n \geq 0$. The relation (5.2) is the **recurrence relation** of the sequence.

The linearly recursive sequence $\{s_n\}$ is **homogeneous** if $a = 0$. The sequence $\{s_n\}$ is **eventually periodic** if there exist integers $N \geq 0$, $t > 0$ for which $s_{n+t} = s_n$, for all $n \geq N$. The sequence $\{s_n\}$ is **periodic** if $\{s_n\}$ is eventually periodic with $N = 0$, that is, $\{s_n\}$ is periodic if there exists an integer $t > 0$ so that $s_{n+t} = s_n$ for all $n \geq 0$. Suppose $\{s_n\}$ is eventually periodic. Then the smallest positive integer r for which $s_{n+r} = s_n$ for all $n \geq N$ is the **period** of $\{s_n\}$.

For $n \geq 0$, the vector $\mathbf{s}_n = (s_n, s_{n+1}, s_{n+2}, \ldots, s_{n+k-1})$ is the **nth state vector** of $\{s_n\}$; $\mathbf{s}_0 = (s_0, s_1, s_2, \ldots, s_{k-1})$ is the **initial state vector**. A linearly recursive sequence is completely determined by specifying the recurrence relation (5.2) and initial state vector.

Here are two basic examples of 1st-order linearly recursive sequences. Let $\mathbf{s}_0 = (s_0)$ for some $s_0 \in K$ and let

$$s_{n+1} = s_n + a$$

for $a \in K$. Then $s_1 = s_0 + a$, $s_2 = s_1 + a = s_0 + 2a$, and so on. The resulting sequence $\{s_n\}$ is the **arithmetic sequence** with initial term s_0 and common difference a. Note that a formula for the nth term of the sequence is $s_n = s_0 + na$. For another example, let $\mathbf{s}_0 = (s_0)$ be the initial state vector and let

$$s_{n+1} = a_0 s_n$$

for some $a_0 \in K$. Now the sequence is s_0, $s_1 = a_0 s_0$, $s_2 = a_0 s_1 = a_0^2 s_0$, and so on. This is the **geometric sequence** with initial term s_0 and ratio a_0. The formula for the nth term of the geometric sequence is $s_n = s_0 a_0^n$.

Perhaps the most well-known 2nd-order linearly recursive sequence is the sequence attributed to Fibonacci. Let $\mathbf{s}_0 = (s_0, s_1)$ be the initial state vector and put

$$s_{n+2} = s_{n+1} + s_n$$

for $n \geq 0$. Then the sequence $\{s_n\}$ is the **Fibonacci sequence**. If $K = \mathbb{Q}$ and the initial state vector is $\mathbf{s}_0 = (0, 1)$, then the Fibonacci sequence is

$$0, 1, 1, 2, 3, 5, 8, 13, 21, \ldots$$

(The Fibonacci sequence appeared in Leonardo of Pisa's Liber Abaci (1202), but was known centuries earlier in India, see [Singh (1985)].)

Homogeneous linearly recursive sequences can be described in terms of matrices. Let $\{s_n\}$ be a homogeneous kth-order linearly recursive sequence with recurrence relation (5.2). Put

$$A = \begin{pmatrix} 0 & 1 & 0 & \cdots & 0 & 0 \\ 0 & 0 & 1 & \cdots & 0 & 0 \\ \vdots & \vdots & \vdots & \ddots & \vdots & \vdots \\ 0 & 0 & 0 & \cdots & 1 & 0 \\ 0 & 0 & 0 & \cdots & 0 & 1 \\ a_0 & a_1 & a_2 & \cdots & a_{k-2} & a_{k-1} \end{pmatrix}.$$

Let M^T denote the transpose of a matrix M.

Proposition 5.12. *With A defined as above,*

$$\mathbf{s}_n^T = A^n \mathbf{s}_0^T$$

for all $n \geq 0$.

Proof. Use induction on n. The trivial case is $n = 0$: $\mathbf{s}_0^T = I_k \mathbf{s}_0^T$. For the induction hypothesis, assume that $\mathbf{s}_{n-1}^T = A^{n-1}\mathbf{s}_0^T$. Then $A\mathbf{s}_{n-1}^T = AA^{n-1}\mathbf{s}_0^T$, hence $\mathbf{s}_n^T = A^n \mathbf{s}_0^T$. □

The matrix A is the **matrix** of the homogeneous linearly recursive sequence. Let $\{s_n\}$ be a homogeneous kth-order linearly recursive sequence with matrix A. The **characteristic polynomial** of $\{s_n\}$ is the characteristic polynomial of A in the usual sense, that is, the characteristic polynomial of $\{s_n\}$ is

$$f(x) = \det(xI_k - A)$$

where I_k denotes the $k \times k$ identity matrix.

Proposition 5.13. *Let $\{s_n\}$ be a homogeneous kth-order linearly recursive sequence defined by (5.2). Let $f(x)$ be the characteristic polynomial of $\{s_n\}$. Then*

$$f(x) = x^k - a_{k-1}x^{k-1} - a_{k-2}x^{k-2} - \cdots - a_0 \in K[x].$$

Proof. We proceed by induction on the order k of the linearly recursive sequence. The trivial case is $k = 1$. The 1st-order linearly recursive sequence $\{s_n\}$ with $s_{n+1} = a_0 s_n$, $n \geq 0$, has associated matrix $A = (a_0)$. Clearly, the characteristic polynomial $f(x) = \det(xI_1 - A) = x - a_0$. Thus the trivial case holds.

For the induction hypothesis, we assume that a $(k-1)$th-order linearly recursive sequence $\{s_n\}$ with matrix

$$B = \begin{pmatrix} 0 & 1 & 0 & \cdots & 0 & 0 \\ 0 & 0 & 1 & \cdots & 0 & 0 \\ \vdots & \vdots & \vdots & \ddots & \vdots & \vdots \\ 0 & 0 & 0 & \cdots & 1 & 0 \\ 0 & 0 & 0 & \cdots & 0 & 1 \\ b_0 & b_1 & b_2 & \cdots & b_{k-3} & b_{k-2} \end{pmatrix}$$

has characteristic polynomial

$$g(x) = \det(xI_{k-1} - B) = x^{k-1} - b_{k-2}x^{k-2} - b_{k-3}x^{k-3} - \cdots - b_0.$$

Now for a given kth-order linearly recursive sequence $\{s_n\}$, one has

$$xI_k - A = \begin{pmatrix} x & -1 & 0 & \cdots & 0 & 0 \\ 0 & x & -1 & \cdots & 0 & 0 \\ \vdots & \vdots & \vdots & \ddots & \vdots & \vdots \\ 0 & 0 & 0 & \cdots & -1 & 0 \\ 0 & 0 & 0 & \cdots & x & -1 \\ -a_0 & -a_1 & -a_2 & \cdots & -a_{k-2} & x - a_{k-1} \end{pmatrix}.$$

Computing $\det(xI_k - A)$ by cofactor expansion about the first column yields

$$\begin{aligned}\det(xI_k - A) &= x \cdot (-1)^{1+1} \cdot \det(M_{1,1}) + (-a_0) \cdot (-1)^{k+1} \cdot (-1)^{k-1} \\ &= x \cdot \det(M_{1,1}) - a_0,\end{aligned}$$

where $M_{1,1}$ is the 1, 1-minor matrix of $xI_k - A$. Of course, $M_{1,1}$ is the $(k-1) \times (k-1)$ matrix of the form $xI_{k-1} - B$ with

$$B = \begin{pmatrix} 0 & 1 & 0 & \cdots & 0 & 0 \\ 0 & 0 & 1 & \cdots & 0 & 0 \\ \vdots & \vdots & \vdots & \ddots & \vdots & \vdots \\ 0 & 0 & 0 & \cdots & 1 & 0 \\ 0 & 0 & 0 & \cdots & 0 & 1 \\ a_1 & a_2 & a_3 & \cdots & a_{k-2} & a_{k-1} \end{pmatrix},$$

and so by the induction hypothesis,

$$\begin{aligned}f(x) &= \det(xI_k - A) \\ &= x \cdot \det(xI_{k-1} - B) - a_0 \\ &= x(x^{k-1} - a_{k-1}x^{k-2} - a_{k-2}x^{k-3} - \cdots - a_1) - a_0 \\ &= x^k - a_{k-1}x^{k-1} - a_{k-2}x^{k-2} - \cdots - a_0.\end{aligned}$$

□

The matrix of the Fibonacci sequence $\{s_n\}$ is

$$A = \begin{pmatrix} 0 & 1 \\ 1 & 1 \end{pmatrix}$$

and the characteristic polynomial is $f(x) = x^2 - x - 1$. From Proposition 5.12, one has

$$\begin{pmatrix} 0 & 1 \\ 1 & 1 \end{pmatrix}^n \begin{pmatrix} s_0 \\ s_1 \end{pmatrix} = \begin{pmatrix} s_n \\ s_{n+1} \end{pmatrix}.$$

Proposition 5.14. *Let K be a field with char$(K) \neq 5$. Then $f(x) = x^2 - x - 1 \in K[x]$ has distinct roots in some extension field L/K.*

Proof. By Proposition 5.2, there exists a field extension L/K so that $f(x)$ factors into linear factors

$$f(x) = (x - \alpha_1)(x - \alpha_2),$$

where α_1, α_2 are the zeros of $f(x)$. Since $\gcd(f(x), f'(x)) = 1$, these zeros are distinct. □

In fact, if α is one zero of $f(x)$, then $1-\alpha$ is the other zero since

$$(1-\alpha)^2 - (1-\alpha) - 1 = 0.$$

Using some elementary linear algebra, we can obtain an explicit formula for the Fibonacci sequence.

Proposition 5.15. *Let K be a field with char$(K) \neq 5$. Let $\{s_n\}$ be the Fibonacci sequence in K with initial state vector $\mathbf{s}_0 = (s_0, s_1)$. Let α be a zero of $f(x) = x^2 - x - 1$ in some extension field L/K. Then*

$$s_n = s_0\left(\frac{\alpha^{n-1}(2\alpha-1) + (1-\alpha)^{n-1}(1-2\alpha)}{5}\right) + s_1\left(\frac{\alpha^{n-1}(2+\alpha) + (1-\alpha)^{n-1}(3-\alpha)}{5}\right),$$

for $n \geq 0$.

Proof. By Proposition 5.14, $f(x)$ has distinct roots, α, $1-\alpha$ in some extension field L/K. Thus by elementary linear algebra, the matrix $\begin{pmatrix} 0 & 1 \\ 1 & 1 \end{pmatrix}$ is diagonalizable. Indeed one obtains,

$$\begin{pmatrix} 0 & 1 \\ 1 & 1 \end{pmatrix} = \begin{pmatrix} \alpha-1 & -\alpha \\ 1 & 1 \end{pmatrix} \begin{pmatrix} \alpha & 0 \\ 0 & 1-\alpha \end{pmatrix} \begin{pmatrix} \frac{-1+2\alpha}{5} & \frac{2+\alpha}{5} \\ \frac{1-2\alpha}{5} & \frac{3-\alpha}{5} \end{pmatrix}.$$

Consequently,

$$\begin{pmatrix} s_n \\ s_{n+1} \end{pmatrix} = \begin{pmatrix} 0 & 1 \\ 1 & 1 \end{pmatrix}^n \begin{pmatrix} s_0 \\ s_1 \end{pmatrix}$$

$$= \begin{pmatrix} \alpha-1 & -\alpha \\ 1 & 1 \end{pmatrix} \begin{pmatrix} \alpha^n & 0 \\ 0 & (1-\alpha)^n \end{pmatrix} \begin{pmatrix} \frac{-1+2\alpha}{5} & \frac{2+\alpha}{5} \\ \frac{1-2\alpha}{5} & \frac{3-\alpha}{5} \end{pmatrix} \begin{pmatrix} s_0 \\ s_1 \end{pmatrix}$$

$$= \begin{pmatrix} \alpha^n(\alpha-1) & -\alpha(1-\alpha)^n \\ \alpha^n & (1-\alpha)^n \end{pmatrix} \begin{pmatrix} \frac{-1+2\alpha}{5} & \frac{2+\alpha}{5} \\ \frac{1-2\alpha}{5} & \frac{3-\alpha}{5} \end{pmatrix} \begin{pmatrix} s_0 \\ s_1 \end{pmatrix}.$$

Thus,

$$s_{n+1} = s_0\left(\frac{\alpha^n(2\alpha-1)+(1-\alpha)^n(1-2\alpha)}{5}\right)$$
$$+ s_1\left(\frac{\alpha^n(2+\alpha)+(1-\alpha)^n(3-\alpha)}{5}\right),$$

or,

$$s_n = s_0\left(\frac{\alpha^{n-1}(2\alpha-1)+(1-\alpha)^{n-1}(1-2\alpha)}{5}\right)$$
$$+ s_1\left(\frac{\alpha^{n-1}(2+\alpha)+(1-\alpha)^{n-1}(3-\alpha)}{5}\right).$$

□

With $\mathbf{s}_0 = (0,1)$, one has

$$s_n = \frac{\alpha^{n-1}(2+\alpha)+(1-\alpha)^{n-1}(3-\alpha)}{5}$$

for $n \geq 0$. In the case that $K = \mathbb{Q}$, $f(x)$ is irreducible over K and we choose $\alpha = \frac{1+\sqrt{5}}{2}$. Thus,

$$\begin{aligned} s_n &= \frac{\alpha^{n-1}(2+\alpha)+(1-\alpha)^{n-1}(3-\alpha)}{5} \\ &= \frac{1}{5}\left(\left(\frac{1+\sqrt{5}}{2}\right)^{n-1}\left(\frac{5+\sqrt{5}}{2}\right)+\left(\frac{1-\sqrt{5}}{2}\right)^{n-1}\left(\frac{5-\sqrt{5}}{2}\right)\right) \\ &= \frac{1}{\sqrt{5}}\left(\left(\frac{1+\sqrt{5}}{2}\right)^{n}-\left(\frac{1-\sqrt{5}}{2}\right)^{n}\right) \end{aligned}$$

for $n \geq 0$.

The Fibonacci sequence over $\mathbb{Q}$ with initial state $\mathbf{s}_0 = (0,1)$ has a surprising connection to the monoid of words built from an alphabet. Let $\{a,b\}$ be an alphabet and let $\{a,b\}^*$ denote the monoid of words of finite length built from the letters in $\{a,b\}$. We consider a subset L of $\{a,b\}^*$ defined as follows: L consists of all words in $\{a,b\}^*$ that contain no consecutive occurrences of b's. For example,

$$e,\ a,\ b,\ aa,\ ab,\ ba,\ aaa,\ aab$$

are words in L; *abaabba* however, is not a word in L.

Proposition 5.16. *Let $\{s_n\}$ denote the Fibonacci sequence over $\mathbb{Q}$ with initial state vector $\mathbf{s}_0 = (0,1)$. Then the number of words in L of length n is s_{n+2} for $n \geq 0$.*

Proof. We use induction on n. Certainly, the claim holds for the trivial cases $n = 0, 1$: for $n = 0$, there is $s_2 = 1$ word of length 0 (the empty word) in L; for $n = 1$, there are $s_3 = 2$ words of length 1 in L, namely a and b. For the induction hypothesis, assume that there are $s_{(n-2)+2} = s_n$ words of length $n-2$ in L and $s_{(n-1)+2} = s_{n+1}$ words of length $n-1$ in L. Let w be a word of length n in L. Then w either ends in a or b. If w ends in a, then it is the concatenation $w = w'a$, where w' is a word of length $n-1$ in L. By the induction hypothesis there are s_{n+1} words of length $n-1$ in L, and so there are s_{n+1} words in L of length n that end in a.

On the other hand, if $w \in L$ ends in b, then the $(n-1)$th letter of w is a, and we have $w = w'ab$, where w' is a word in L of length $n-2$. By the induction hypothesis there are s_n words in L of length $n-2$, thus there are s_n words in L that end in b. It follows that there are $s_{n+2} = s_{n+1} + s_n$ words in L of length n. □

Let K be a field, let

$$g(x) = b_l x^l + b_{l-1}x^{l-1} + \cdots + b_1 x + b_0$$

be a polynomial in $K[x]$ and let A be a matrix in $\mathrm{Mat}_k(K)$. The evaluation of $g(x)$ at A, denoted as $g(A)$, is the linear combination of matrices

$$g(A) = b_l A^l + b_{l-1}A^{l-1} + \cdots + b_1 A + b_0 I_k.$$

The polynomial $g(x)$ **annihilates** A if $g(A) = 0$, where 0 denotes the $k \times k$ zero matrix. The set of all polynomials in $K[x]$ that annihilate A is a non-zero ideal J of $K[x]$ (§5.4, Exercise 16). Since $K[x]$ is a PID, $J = (p(x))$, for some monic polynomial $m(x)$. The polynomial $m(x)$ is called the **minimal polynomial of** A.

The **Cayley-Hamilton Theorem** of linear algebra states that the minimal polynomial of A divides the characteristic polynomial of A, see [Hoffman and Kunze (1971), §7.1]. If A has the form of the matrix of a homogeneous linearly recursive sequence, then we can say a bit more.

Proposition 5.17. *Let A be a matrix in $Mat_k(K)$ of the form*

$$A = \begin{pmatrix} 0 & 1 & 0 & \cdots & 0 & 0 \\ 0 & 0 & 1 & \cdots & 0 & 0 \\ \vdots & \vdots & \vdots & \ddots & \vdots & \vdots \\ 0 & 0 & 0 & \cdots & 1 & 0 \\ 0 & 0 & 0 & \cdots & 0 & 1 \\ a_0 & a_1 & a_2 & \cdots & a_{k-2} & a_{k-1} \end{pmatrix}.$$

Then the characteristic polynomial of A is the minimal polynomial of A.

Proof. The proof of this well-known result uses the Cayley-Hamilton Theorem, see [Hoffman and Kunze (1971), §7.1, Corollary]. □

An application of Propositon 5.17 says that the characteristic polynomial of a linearly recursive sequence is the minimal polynomial of its matrix.

For the remainder of this section we are going to consider kth-order homogeneous linearly recursive sequences $\{s_n\}$ in the field $K = \mathrm{GF}(p^m)$, the Galois field of p^m elements. We will assume, without explicitly stating it, that our linearly recursive sequences are homogeneous. The fundamental result on kth-order linearly recursive sequences in $\mathrm{GF}(p^m)$ is the following: if the characteristic polynomial of the sequence is primitive, then the sequence is periodic with maximal period $r = p^{mk} - 1$. This makes linearly recursive sequences over finite fields useful for the generation of pseudorandom strings, such as the shrinking generator sequence, see [Menezes *et al.* (1997)]. One thing we notice right away is that every linearly recursive sequence in a finite field is eventually periodic.

Proposition 5.18. *Let $\{s_n\}$ be a kth-order linearly recursive sequence in GF(p^m). Then $\{s_n\}$ is eventually periodic with period $1 \le r \le p^{mk} - 1$.*

Proof. If $\mathbf{s}_i = (\underbrace{0, 0, \ldots, 0}_{k})$ for any $i \ge 0$, then $\{s_n\}$ is eventually periodic with $N = i$ and $r = 1 \le p^{mk} - 1$. So we assume that $\mathbf{s}_n$ is not the zero vector for all $n \ge 0$.

Observe that there are precisely $p^{mk} - 1$ distinct k-tuples of elements in $\mathrm{GF}(p^m)$ other than the zero vector. Thus there exist integers i, j, $0 \le i < j \le p^{mk} - 1$, so that $\mathbf{s}_j = \mathbf{s}_i$. We claim that $\mathbf{s}_{n+j-i} = \mathbf{s}_n$ for all $n \ge i$. To prove this claim we proceed by induction on $n \ge i$, with the trivial case $n = i$ already established. For the induction hypothesis, we assume that $\mathbf{s}_{n+j-i} = \mathbf{s}_n$ holds for $n = i + \omega$, where $\omega \ge 0$ is a fixed integer. Thus $\mathbf{s}_{\omega+j} = \mathbf{s}_{i+\omega}$, and hence

$$(s_{\omega+j}, s_{\omega+j+1}, \ldots, s_{\omega+j+k-1}) = (s_{i+\omega}, s_{i+\omega+1}, \ldots, s_{i+\omega+k-1}). \quad (5.3)$$

Consequently

$$s_{\omega+1+j+\eta} = s_{i+\omega+1+\eta},$$

for $\eta = 0, 1, 2, \ldots, k-2$. Moreover, from (5.3) and the recurrence relation (5.2) one obtains

$$s_{\omega+1+j+\eta} = s_{i+\omega+1+\eta},$$

for $\eta = k - 1$. Thus

$$\begin{aligned}(s_{(\omega+1)+j}, s_{(\omega+1)+j+1}, \ldots, s_{(\omega+1)+j+k-1})\\ = (s_{i+(\omega+1)}, s_{i+(\omega+1)+1}, \ldots, s_{i+(\omega+1)+k-1}),\end{aligned}$$

which yields $\mathbf{s}_{n+j-i} = \mathbf{s}_n$ for $n = i + (\omega + 1)$. The proof by induction is complete and hence $s_{n+(j-i)} = s_n$ for all $n \geq i$. Thus$\{s_n\}$ is eventually periodic with $N = i$ and period r satisfying $1 \leq r \leq j - i \leq p^{mk} - 1$. □

Proposition 5.19. *Let $\{s_n\}$ be a kth-order linearly recursive sequence in GF(p^m) with recurrence relation (5.2). If $a_0 \neq 0$, then $\{s_n\}$ is periodic with period $1 \leq r \leq p^{mk} - 1$.*

Proof. By Proposition 5.18, $\{s_n\}$ is eventually periodic with period $1 \leq r \leq p^{mk} - 1$. Suppose N_0 is the smallest integer for which $s_{n+r} = s_n$ for all $n \geq N_0$. If $N_0 = 0$ then $\{s_n\}$ is periodic. So we assume that $N_0 \geq 1$. But now from (5.2),

$$\begin{aligned}s_{N_0-1+r} &= a_0^{-1}\left(s_{(N_0-1+r)+k} - a_{k-1}s_{(N_0-1+r)+k-1} - \cdots - a_1 s_{(N_0-1+r)+1}\right)\\ &= a_0^{-1}\left(s_{N_0+k-1+r} - a_{k-1}s_{N_0+k-2+r} - \cdots - a_1 s_{N_0+r}\right)\\ &= a_0^{-1}\left(s_{N_0+k-1} - a_{k-1}s_{N_0+k-2} - \cdots - a_1 s_{N_0}\right)\\ &= a_0^{-1}\left(s_{(N_0-1)+k} - a_{k-1}s_{(N_0-1)+k-1} - \cdots - a_1 s_{(N_0-1)+1}\right)\\ &= s_{N_0-1}.\end{aligned}$$

Thus $s_{n+r} = s_n$ for all $n \geq N_0 - 1$, which contradicts the minimality of N_0. □

Let $GL_k(\mathrm{GF}(p^m))$ denote the group of invertible $k \times k$ matrices with entries in GF(p^m). By [Hoffman and Kunze (1971), §5.4, Theorem 4], a $k \times k$ matrix A over GF(p^m) is invertible if and only if $\det(A) \neq 0$. Let $\{s_n\}$ be a kth-order linearly recursive sequence in GF(p^m) with recurrence relation (5.2). In what follows we assume that $a_0 \neq 0$ so that $\{s_n\}$ is periodic with period r, $1 \leq r \leq p^{mk} - 1$. Let

$$A = \begin{pmatrix} 0 & 1 & 0 & \cdots & 0 & 0\\ 0 & 0 & 1 & \cdots & 0 & 0\\ \vdots & \vdots & \vdots & \ddots & \vdots & \vdots\\ 0 & 0 & 0 & \cdots & 1 & 0\\ 0 & 0 & 0 & \cdots & 0 & 1\\ a_0 & a_1 & a_2 & \cdots & a_{k-2} & a_{k-1}\end{pmatrix}$$

be the matrix of $\{s_n\}$. Observe that $\det(A) = (-1)^{k+1}a_0$ and consequently, $A \in GL_k(\mathrm{GF}(p^m))$, a finite group. Hence A has finite order in $GL_k(\mathrm{GF}(p^m))$.

Proposition 5.20. *Let $\{s_n\}$ be a kth-order linearly recursive sequence in GF(p^m). Then the period r of $\{s_n\}$ divides the order of A in $GL_k(GF(p^m))$.*

Proof. Let l be the order of A. Then from Proposition 5.12,

$$\mathbf{s}_{n+l}^T = A^{n+l}\mathbf{s}_0^T = A^n\mathbf{s}_0^T = \mathbf{s}_n^T,$$

for all $n \geq 0$. Thus $r \leq l$. There exist integers t, u so that $l = rt + u$ with $0 \leq u < r$, $t > 0$. Now, for all $n \geq 0$,

$$\mathbf{s}_n^T = \mathbf{s}_{n+l}^T = \mathbf{s}_{n+rt+u}^T = \mathbf{s}_{n+r+r(t-1)+u}^T = \mathbf{s}_{n+r(t-1)+u}^T,$$

$$\mathbf{s}_{n+r(t-1)+u}^T = \mathbf{s}_{n+r+r(t-2)+u}^T = \mathbf{s}_{n+r(t-2)+u}^T,$$

$$\mathbf{s}_{n+r(t-2)+u}^T = \mathbf{s}_{n+r+r(t-3)+u}^T = \mathbf{s}_{n+r(t-3)+u}^T,$$

$$\vdots$$

$$\mathbf{s}_{n+r+u}^T = \mathbf{s}_{n+u}^T.$$

Consequently, $\mathbf{s}_n^T = \mathbf{s}_{n+u}^T$. Since r is the period of $\{s_n\}$, $u = 0$, and so $r \mid l$. □

Let

$$f(x) = x^k - a_{k-1}x^{k-1} - a_{k-2}x^{k-2} - \cdots - a_0 \in \mathrm{GF}(p^m)[x]$$

be the characteristic polynomial of A. Here is a technical proposition that we need soon.

Proposition 5.21. *Let $\{s_n\}$ be a kth-order linearly recursive sequence in GF(p^m)[x] with characteristic polynomial $f(x)$, $a_0 \neq 0$. Let r be the period of $\{s_n\}$. Let*

$$v(x) = s_0x^{r-1} + s_1x^{r-2} + \cdots + s_{r-2}x + s_{r-1},$$

$$w(x) = \sum_{j=0}^{k-1}(a_{j+1}s_0 + a_{j+2}s_1 + a_{j+3}s_2 + \cdots + a_{k-1}s_{k-2-j} + a_ks_{k-1-j})x^j,$$

$$\textit{with} \quad a_k = -1.$$

Then

$$f(x)v(x) = (1 - x^r)w(x).$$

Proof. Using the recurrence relation

$$s_{n+k} = a_{k-1}s_{n+k-1} + a_{k-2}s_{n+k-2} + \cdots + a_0 s_n$$

and the formula $s_{n+r} = s_n$, valid for $n \geq 0$, one shows directly that terms of $f(x)v(x)$ coincide with those of $(1 - x^r)w(x)$. □

Proposition 5.22. *Let $\{s_n\}$ be a kth-order linearly recursive sequence in GF(p^m). Let A be the matrix of $\{s_n\}$ and let $f(x)$ be the characteristic polynomial of A. Assume that $a_0 \neq 0$. Then order($f(x)$) equals the order of the matrix A in the finite group $GL_k(GF(p^m))$.*

Proof. The order of A is the smallest positive integer l so that $A^l - I_k = 0$. That is, l is the smallest positive integer so that $x^l - 1$ is in the annihilator ideal of A. Consequently, the minimal polynomial $m(x) \mid x^l - 1$. Since $f(x) = m(x)$ by Proposition 5.17, $f(x) \mid x^l - 1$. Thus order$(f(x)) \leq l$. If order$(f(x)) < l$, then there exists an integer q, $0 < q < l$, with $f(x) \mid x^q - 1$. Consequently, $A^q - I_k = 0$, which contradicts our assumption that l is the order of A. It follows that order$(f(x)) = l$. □

Proposition 5.23. *Let $\{s_n\}$ be a kth-order linearly recursive sequence in GF(p^m). Let A be the matrix of $\{s_n\}$ and let $f(x)$ be the characteristic polynomial of A. Assume that $a_0 \neq 0$. Let r be the period of $\{s_n\}$. Then $r \mid$ order($f(x)$).*

Proof. Let l be the order of A in GL_k(GF(p^m)). By Proposition 5.20, $r \mid l$. By Proposition 5.22, $l =$ order$(f(x))$. Thus $r \mid$ order$(f(x))$. □

Proposition 5.24. *Let $\{s_n\}$ be a kth-order linearly recursive sequence in GF(p^m). Let A be the matrix of $\{s_n\}$ and let $f(x)$ be the characteristic polynomial of A. Assume that $a_0 \neq 0$. Let r be the period of $\{s_n\}$. If $f(x)$ is irreducible over GF(p^m), then $r =$ order($f(x)$).*

Proof. By Proposition 5.23, $r \mid$ order$(f(x))$. Thus $r \leq$ order$(f(x))$. By Proposition 5.21, $f(x) \mid (1 - x^r)w(x)$ for some $w(x) \in$ GF$(p^m)[x]$, $w(x) \neq 0$. Since $f(x)$ is irreducible, either $f(x) \mid 1 - x^r$ or $f(x) \mid w(x)$. Since $\deg(w(x)) < \deg(f(x))$, one has $f(x) \mid 1 - x^r$ and so, order$(f(x)) \leq r$. □

Here is the key result regarding linearly recursive sequences.

Proposition 5.25. *Let $\{s_n\}$ be a kth-order linearly recursive sequence in GF(p^m) with characteristic polynomial $f(x)$. Assume that $a_0 \neq 0$ and let r be the period of $\{s_n\}$. If $f(x)$ is primitive over GF(p^m), then $r = p^{mk} - 1$.*

Proof. By Proposition 5.24, $r = \text{order}(f(x))$. Since $f(x)$ is primitive of degree k, $\text{order}(f(x)) = p^{mk} - 1$. □

As we have seen, $f(x) = x^4 + x + 1$ is a primitive polynomial over GF(2). Thus we can apply Proposition 5.25 to produce a 4th-order linearly recursive sequence $\{s_n\}$ in GF(2) that has maximal period $r = 2^4 - 1 = 15$. From $f(x) = x^4 + x + 1$ we obtain the recurrence relation

$$s_{n+4} = s_{n+1} + s_n, \quad n \geq 0.$$

Choosing the initial state vector $\mathbf{s}_0 = 0110$, we obtain the sequence

$$011010111100010011010111100010011\ldots$$

of maximal period 15. As another example, the initial state vector $\mathbf{s}_0 =$ 0001 yields the sequence

$$000100110101111000100110101111000\ldots$$

of period 15. If the initial 4-tuple $\mathbf{s}_0$ is chosen randomly (by a coin flip, perhaps) then we can extend the random sequence $\mathbf{s}_0$ to a pseudorandom sequence of length 15.

Here is another example of a pseudorandom sequence of bits. As the base field, choose $\text{GF}(16) = \text{GF}(2)(\alpha)$, where $\alpha^4 + \alpha + 1 = 0$. Let $f(x) = x^3 + \alpha x^2 + \alpha^2 \in \text{GF}(16)[x]$. We ask: Is $f(x)$ irreducible over GF(16)? Is $f(x)$ primitive over GF(16)? At any rate, we can consider the 3rd-order linearly recursive sequence $\{s_n\}$ in GF(16) whose recurrence relation is

$$s_{n+3} = \alpha s_{n+2} + \alpha^2 s_n, \quad n \geq 0.$$

The period r of $\{s_n\}$ satisfies $1 \leq r \leq 2^{12} - 1 = 4095$. Writing elements of GF(16) as half-bytes (as in Example 5.4) we see that

$$s_{n+3} = 0100 s_{n+2} + 0010 s_n, \quad n \geq 0.$$

We let $\mathbf{s}_0 = 1010\ 0111\ 1000$ where the 0's and 1's are chosen uniformly at random from $\{0, 1\}$. What does the sequence look like? We have

$$s_3 = 0100 \cdot 1000 + 0010 \cdot 1010 = 1010,$$

so the sequence begins

$$1010\ 0111\ 1000\ 1010\ldots$$

5.4 Exercises

Exercises for §5.1

(1) Prove that $f(x) = x^3 + 2x^2 + x + 1$ is irreducible over Z_3. Find an invented root of $f(x)$.

(2) Let F be a field, let $f(x) \in F[x]$, and let $g(x) \in (f(x))$. Show that $f(x + g(x)) \in (f(x))$.

Exercises for §5.2

(3) Let GF(8) denote the finite field of $2^3 = 8$ elements.

 (a) Factor the polynomial $x^8 - x$ into a product of irreducible polynomials over Z_2.
 (b) Using invented roots, write GF(8) as a simple algebraic extension of Z_2.
 (c) Using part (b) write each element of GF(8) as a sequence of 3 bits.
 (d) Using parts (b) and (c) compute $011 \cdot 101$ in GF(8).

(4) Let GF(9), GF(81) be the Galois fields with 9, 81 elements, respectively. Find an irreducible polynomial $f(x) \in \mathrm{GF}(9)[x]$ and a root β of $f(x)$ so that $\mathrm{GF}(81) = \mathrm{GF}(9)(\beta)$.

(5) Let $n \geq 1$ be an integer and let $Z_p = \mathrm{GF}(p)$ denote the Galois field with p elements. Prove that there exists an irreducible polynomial of degree n over Z_p.

(6) Determine whether $f(x) = x^4 + x^3 + 1$ is a primitive polynomial over GF(2).

(7) Determine whether $f(x) = x^3 + 2x^2 + 1$ is a primitive polynomial over GF(3).

Exercises for §5.3

(8) Let $\{s_n\}$ be the sequence in $\mathbb{Q}$ given as

$$1, 2, 3, 4, 1, 2, 3, 4, 1, 2, 3, 4, 1, 2, 3, 4, \ldots.$$

Show that $\{s_n\}$ is a linearly recursive sequence. What is its initial state vector? What is its recurrence relation?

(9) Let $\{s_n\}$ be the sequence in $\mathbb{Q}$ given as

$$1, 2, 3, 4, 5, 6, 7, 8, \ldots.$$

Show that $\{s_n\}$ is a 3rd-order linearly recursive sequence.

(10) Let $\{s_n\}$ be the sequence in $\mathbb{Q}$ given as:

$$3, 2, 5, 1, 7, 0, 0, 0, 0, 0 \ldots.$$

Show that $\{s_n\}$ is a linearly recursive sequence. What is its initial state vector? What is its recurrence relation?

(11) Let $\{s_n\}$ be the sequence in $\mathbb{Q}$ given as:

$$1, 2, 2, 3, 3, 3, 4, 4, 4, 4, 5, 5, 5, 5, 5, 6, 6, 6, 6, 6, 6, \ldots.$$

Show that $\{s_n\}$ is not a linearly recursive sequence in $\mathbb{Q}$.

(12) Let $\{s_n\}$ be the 2nd-order linearly recursive sequence in $\mathbb{R}$ with recurrence relation

$$s_{n+2} = 2s_{n+1} + s_n$$

and intital state vector $\mathbf{s}_0 = (-1, 2)$.

(a) Compute the matrix and the characteristic polynomial for $\{s_n\}$.
(b) Find a formula for the nth term of the sequence.

(13) Let $L \subseteq \{a, b\}^*$ be the language consisting of all words that contain no consecutive occurrences of b's. How many words of length 23 are in L?

(14) Let GF(5) be the Galois field with 5 elements. Write out the first five terms of the linearly recursive sequence $\{s_n\}$ with recurrence relation

$$s_{n+2} = 4s_{n+1} + s_n,$$

and initial state vector $\mathbf{s}_0 = (2, 3)$. What is the period of $\{s_n\}$?

(15) Let $\{s_n\}$ be the Fibonacci sequence in GF(5).

(a) Compute the period r of $\{s_n\}$.
(b) Verify that $f(x)v(x) = (1 - x^r)w(x)$ as in Proposition 5.21.

(16) Let L be any field and let $A \in \text{Mat}_k(L)$. Prove that the set of polynomials in $L[x]$ that annihilate A is a non-zero ideal of $L[x]$.

Questions for Further Study

(1) Let $\text{GF}(9) = \text{GF}(3)(\alpha)$ be the Galois field with 9 elements (Example 5.2). Write out the first six terms of the linearly recursive sequence $\{s_n\}$ with recurrence relation

$$s_{n+5} = s_{n+3} + s_{n+1},$$

and initial state vector $\mathbf{s}_0 = (1, \alpha, 2, 1, \alpha)$. What is the period of $\{s_n\}$?

(2) Let GF(16) $=$ GF(2)(α), $\alpha^4 + \alpha + 1 = 0$, denote the Galois field with 16 elements.

(a) Determine whether $f(x) = x^3 + \alpha x^2 + \alpha^2$ is irreducible over GF(16). Is $f(x)$ primitive over GF(16)?

(b) Let $\{s_n\}$ be the 3rd-order linearly recursive sequence in GF(16) with recurrence relation

$$s_{n+3} = \alpha s_{n+2} + \alpha^2 s_n$$

and initial state vector $\mathbf{s}_0 =$ 1010 0111 1000. Compute the first 8 terms of $\{s_n\}$. What is the period of $\{s_n\}$? Note: 1010, for instance, is the coordinate vector of the element $1 + \alpha^2 \in$ GF(16) with respect to the GF(2)-basis $\{1, \alpha, \alpha^2, \alpha^3\}$.

Bibliography

Cassels, J. W. S. and Fröhlich, A. (1967). *Algebraic Number Theory* (Academic Press, London).

Higman, G. (1940). The Units of group-rings, *Proc. London Math. Soc.* **46**, 2, pp. 231–248.

Hoffman, K. and Kunze, R. (1971). *Linear Algebra*, 2nd edn. (Prentice-Hall, New Jersey).

Ireland, K. and Rosen, M. (1990). *A Classical Introduction to Modern Number Theory* (Springer-Verlag, New York).

Lang, S. (1984). *Algebra*, 2nd edn. (Addison-Wesley, Reading, Mass.).

Mao, W. (2004). *Modern Cryptography* (Prentice Hall PTR, New Jersey).

Martin, G. E. (1982). *Transformation Geometry* (Springer-Verlag, New York).

Menezes, A. J., van Oorschot, P. C., and Vanstone, S. A. (1997). *Handbook of Applied Cryptography* (CRC Press, Boca Raton, Fla.).

Neukirch, J. (1999). *Algebraic Number Theory*, Vol. 322 (Springer-Verlag).

Rotman, J. (2002). *Advanced Modern Algebra* (Pearson, New Jersey).

Samuel, P. (2008). *Algebraic Theory of Numbers* (Dover, New York).

Sehgal, S. K. (2013). Units of integral group rings–a survey, `www.math.ualberta.ca/people/Faculty/Sehgal/publications/057.pdf`.

Serre, J.-P. (1977). *Linear Representations of Finite Groups* (Springer-Verlag, New York).

Singh, P. (1985). The So-called fibonacci numbers in ancient and medieval India, *Hist. Math.* **12**, 3, pp. 229–244.

Sloane, N. J. A. (2015). Online encyclopedia of integer sequences, `www.oeis.org`.

Stark, H. (1967). There is no tenth complex quadratic field with class-number one, *Proc. Nat. Acad. Sci. U.S.A.* **57**, pp. 216–221.

Underwood, R. G. (2011). *An Introduction to Hopf Algebras* (Springer, New York).

Washington, L. C. (1997). *Introduction to Cyclotomic fields* (Springer-Verlag, New York).

Index

Printed in the United States
By Bookmasters